Arkansas Sky Observatories'

SUPERNOVA SEARCH ATLAS and GUIDE

Complete CCD Atlas With Research Study Program

P. Clay Sherrod

Revision 1

Printed in the United States of America
 Published 2018
Lulu Press

Arkansas Sky Observatories Publications
Available from: http://arksky.org/publications
Also available on-line as PDF
ISBN 978-1-387-53719-8
Revision 1 – May 14, 2018

Arkansas Sky Observatories'

SUPERNOVA SEARCH ATLAS and GUIDE

Complete CCD Atlas With Research Study Program

by

P. Clay Sherrod
Arkansas Sky Observatories
IAU/MPC H41, H43, H45

Supernova in Messier 51, June 26, 2011
Arkansas Sky Observatories

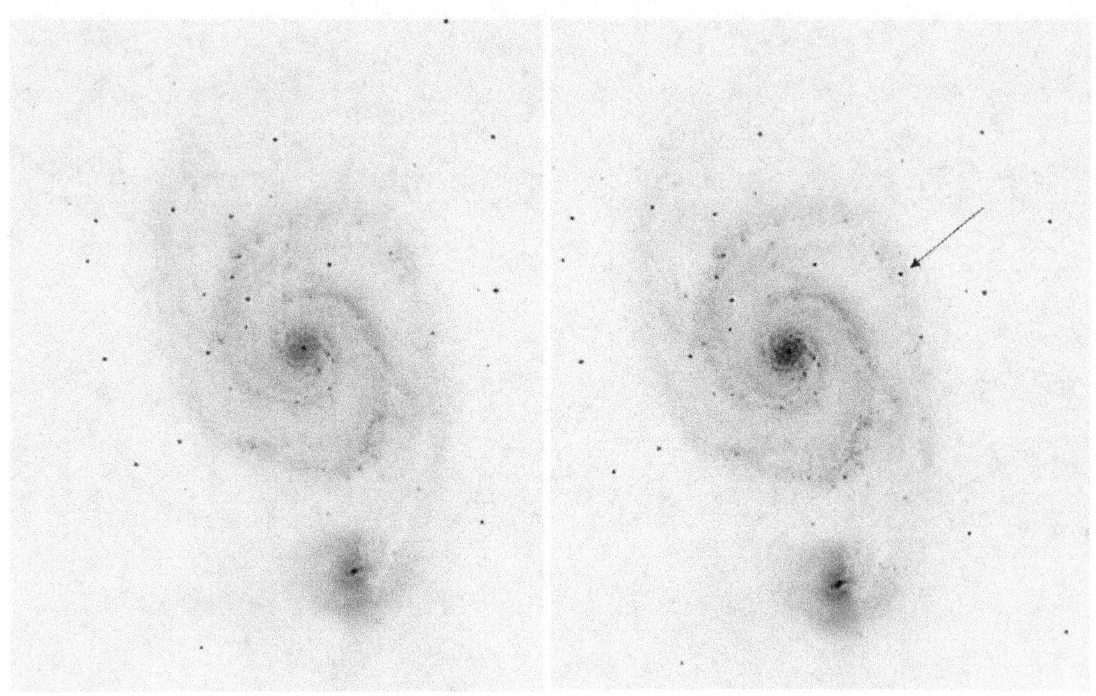

Compiled from over 300 Galaxies in the NGC Catalog

Using the exclusive ASO index and CCD images of over 300 galaxy survey candidates
From the *New General Catalog* (NGC)

Introduction

The study of external galaxies for the explosive supernovae (stars undergoing cataclysmic demise) is not only a vital requirement for our knowledge of stellar evolution, but a convenient and exciting undertaking which can provide both advanced notification of such events as well as an exciting contribution to the field of astronomy.

Too often many sophisticated non-professional and institutional (i.e., school/university) astronomical observatories are under-utilized in terms of scientific study. Outside of curriculum requirements and public awareness events, small observatories throughout the world sit virtually unused for scientific pursuits, when in reality the advancement of computerized and digital imaging allows such facilities vast opportunities to contribute to the field of astronomy and astrophysics.

Among those studies are both astrometry (study of position and motion) of Near Earth Objects (asteroids/minor planets with the potential for impacting Earth), photometry (the study of the brightness changes and rotational possibilities of objects), and cataclysmic stars which exhibit variable activity or in more rare cases destructive events such as **novae** [1] (within our own Milky Way Galaxy) and the even more rare **supernovae** [2] in external galaxies.

Although the processes of the novae and supernovae are somewhat alike and their studies quite similar, the rarity of the supernovae event makes our understanding and study of them far more limited. Novae, on the other hand, are observed yearly throughout our own galaxy and the events are far more understood than the supernovae process. A supernova in our own Milky Way would be a quite rare event, the last taking place in **1604**, [3] but perhaps the most dramatic being that of **1054** [4] A.D., the result of which was the intensely studied Crab Nebula in Taurus.

However, there is a very convenient method through which astronomers can study the supernovae events and processes rather than waiting centuries for one to occur nearby – simply examine other galaxies for events within their realms and be ready when such events occur, because early moments of the events are the most crucial in terms of our understanding of each star's demise.

There are literally billions of other "external galaxies" throughout space, most of which are at distances that render the study of even the brightest of explosive events. But there are millions within reach of telescopes that can reveal supernovae events, and of those there are about a thousand that can be studied with the aforementioned non-professional modern equipment.

Of the millions within study range, we must eliminate those galaxies that are not suitable or more appropriately *less likely* for the outburst of a supernovae. The following restrictions are recommended and used in this study:

1. galaxy must be spiral galaxy, **not elliptical** [5] for likelihood of supernovae
2. target must be close enough for stellar explosion to be visible from Earth
3. target should be large/close enough for supernovae magnitude recordable with limitations of personal/institutional equipment.

Thus, of the thousands that actually are capable of exhibiting a supernova event for modern, modest telescopes and CCD equipment, I have selected what I consider the "**top 300**" galaxies and studied their surroundings to assist in individual compilations of a study plan for supernovae discovery, measuring, recording and reporting.

In addition to detailed guides in terms of setting up an organized and progressive study and research that may lead to numerous supernovae discoveries, this Guide and Atlas provides a considerable degree of standardization in search and research procedures.

This Atlas contains the best 300-plus spiral and irregular galaxies throughout the span of the night sky that will most likely result in reward for your efforts in supernovae searches. Using the precision astrometric 0.51m ASO Astrograph, which operates at f/5.9 with cooled CCD camera, images of these galaxies were obtained from 2017 through 2018 all of which are carefully calibrated in terms of orientation and density; this ATLAS is found after the Study discussion. This telescope, located at IAU station H45, Petit Jean Mountain South (Arkansas), is robotically drive via an extremely accurate Astro-Physics AP1600 mount with absolute encoders, controlled by Project Pluto's *GUIDE* sky program.

Each image obtained was taken under a dark and moonless night with as best sky conditions as possible. Galaxies to 17th magnitude are recorded, but only those brighter than 15th magnitude are suggested for searches. Stellar magnitude reached (depending on altitude from the horizon, and refraction) is between 19 and 19.5, each obtained with a 60-second exposure. All images are identically oriented, with north up and east to the left, with a scale of 22 arc minutes wide by 15 arc minutes top to bottom. All images are RAW and unprocessed except for brightness and converted from original .FITS format to publication-low-resolution quality .jpg format.

Southern objects lower than declination –26 are greatly affected by atmospheric distortion due to local geographical conditions of the southern brow of Petit Jean Mountain and thus are not as sharp as desired. No objects below –29 degrees declination are included in this Atlas.

Edge-on galaxy NGC 891 – Arkansas Sky Observatories photograph

<u>**A Brief Discussion of Supernovae and the Supernova Process**</u>

Some discussion of the processes and characteristics of supernovae is necessary to understand the urgency of early detection and subsequent follow-up on these objects. Complete volumes exist, even on specific events and an excellent summary and reference is found on Wikipedia.org [6]: https://en.wikipedia.org/wiki/History_of_supernova_observation

To study the supernovae experience requires as much as a year of monitoring the progress of the event, carefully measuring the light of the star as it peaks in brightness and then dims, either quite slowly or possibly rapidly.

The first type of supernova that was classified is a binary system, or **Supernova Type 1-a**, in which two stars are orbiting closely, and one of them is a very dense white dwarf star [7]. Without going too far away from this Study Guide discussion, the process is believed to be an increasing flow of matter and energy from a larger star into the more dense and massive white dwarf star which is gravitationally pulling material from the parent star in an ever-increasing manner. Once the star attains a critical mass of 1.44 mass of the original white dwarf, the star collapses under the weight of the accreted mass and the fusion results in a neutron star of considerable mass.

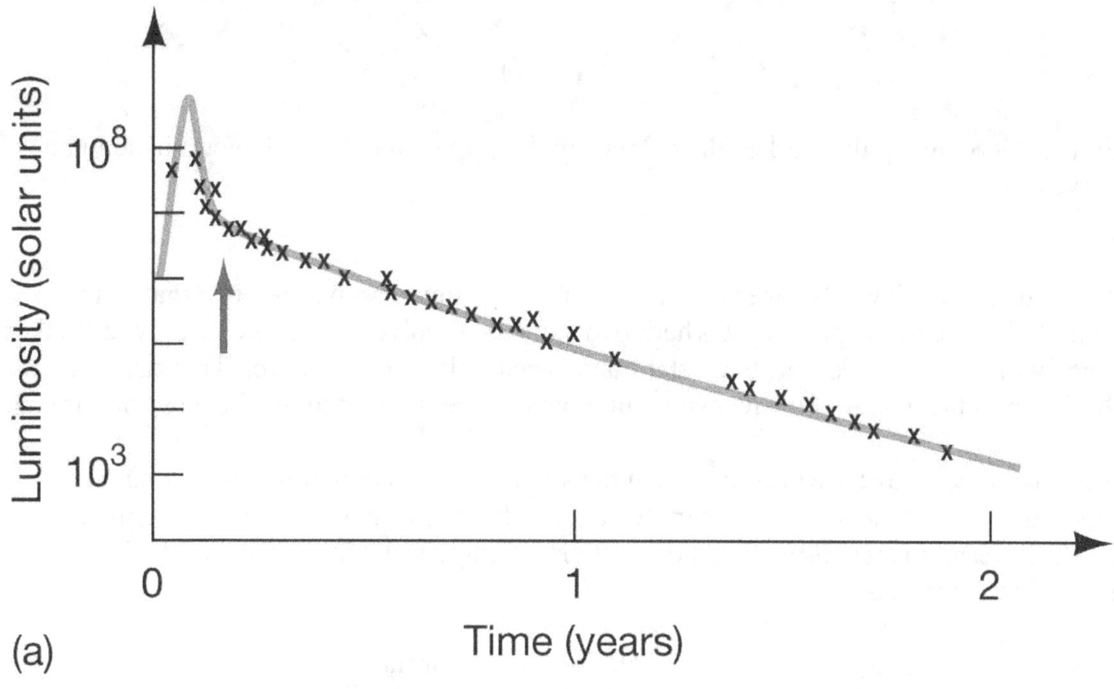

Figure 1 – Light curve of typical Type 1a Supernova. Courtesy Univ. of Oregon [8]

Notice from **Figure 1** the very rapid rise in luminosity (brightness or magnitude in your CCD), the very short and pronounced peak, and then equally rapid decline in brightness by a couple of magnitudes. After that sudden drop, which can be only weeks are even days, the magnitude decline is marked by a very linear and slow decline as can be seen after the arrow pointing to the stabilization period of the stars involved.

Less common are the Type II supernovae which characteristically are easily identified not only through the spectrum that they exhibit, but by the light curve that they demonstrate after lengthy study of their brightness over the course of a couple of years.

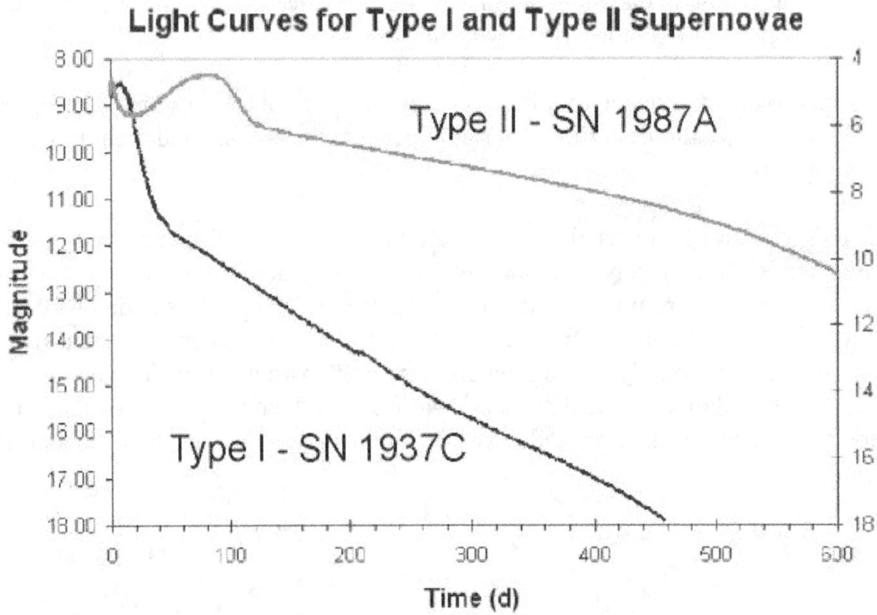

Figure 2 – Showing the marked differences in the Type I and Type II supernovae light curves

---

The nature of the Type II supernova is actually a bit more captivating than that of the Type I. Unlike the Type I event where two stars are involved, the more rare Type II takes place with only a single star, but a star that exceedingly more massive. This star can be 40-50 times the mass of our own sun, but must be at least eight times the mass of our sun.

The supernova II event will occur when the star is so massive that it can no longer thermally expend the force necessary to support the mass and the star rapidly collapses in on itself, result in an enormous fusion event releasing considerable energy which is formed in the process.

It is believed that anything larger than 50 times the solar mass would result in the formation of a black hole, rather than a visible supernova event.

Note the considerable differences in the light curves of the Type I and Type II supernovae. Unlike the immediate decline in brightness of the Type I, the Type II event sustains its brightness over a long period, many times up to 2-3 years after the explosion. This plateau of brightness can be marked by both slow and rapid fluctuations in luminosity.

Thus, even without a spectral analysis available to the modest observatory, the light curve of the supernova event is enough – like a fingerprint – to determine between exactly what type of event occurred and what type of star(s) triggered the supernova.

It is strongly suggested that all observatories not only participate in supernovae searches, but also in the development of all-important magnitude and light curve analysis.

There are many X-Y plotting programs available, the X-axis always being elapsed time, while the Y axis should always specify magnitude or luminosity, the brightest point being the highest on the Y axis.

One such program which can handle both batch as well as individual observations of magnitudes is **VStar** from the American Association of Variable Star Observers (AAVSO) [9]: https://www.aavso.org/vstar

Also available on the AAVSO website is the very convenient Light Curve Generator which requires no download and can be used for rapid acquisition of light curve plots: https://www.aavso.org/lcg

An excellent paper on plotting supernovae light curves has been generated by MIT [10] which not only discusses the ease at which plots can be made, but also delves into the world of UBVRI color photometry, using special filters to greatly improve our understanding of the supernova process through differential spectrography: http://web.mit.edu/asf/www/irlab/Supernova/PhotometryPrimer5.pdf

However, for the purposes of this study, we shall focus on discovery and follow-up, with a brief emphasis on only one part of the UBVRI spectrum, that being the "V" band of the spectrum in which the most accurate and standardization of photometry (the measurement of brightness/luminosity) has been established.

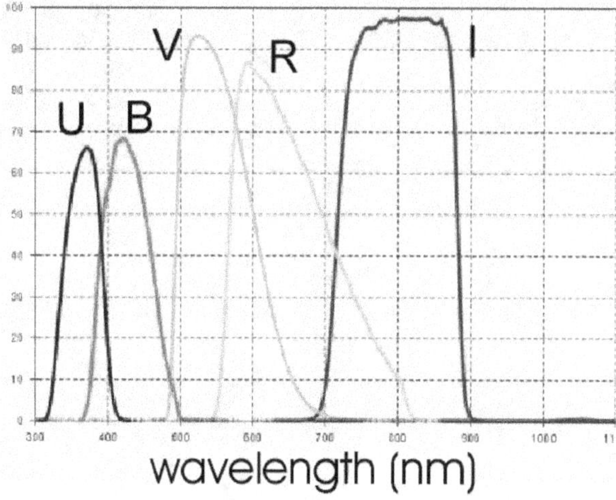

Figure 3 – UBVRI Spectral Transmissions

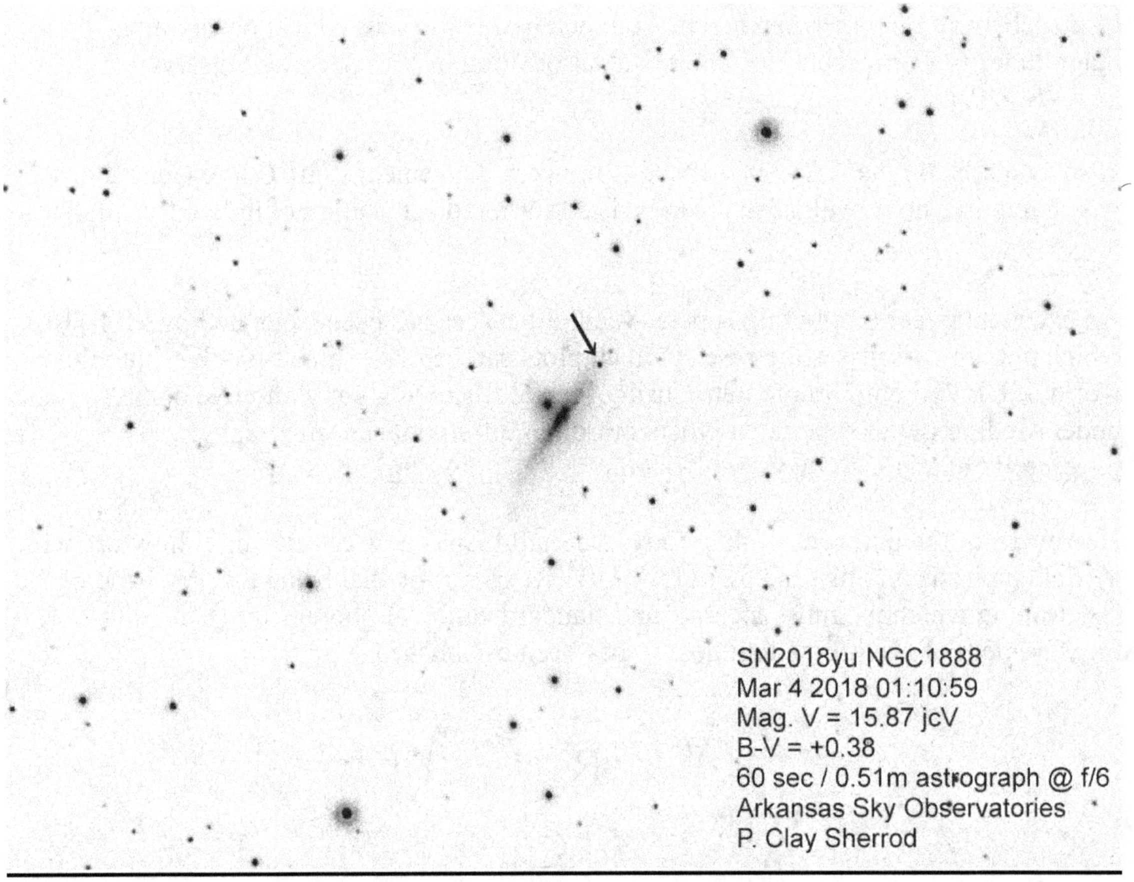

SN2018yu NGC1888
Mar 4 2018 01:10:59
Mag. V = 15.87 jcV
B-V = +0.38
60 sec / 0.51m astrograph @ f/6
Arkansas Sky Observatories
P. Clay Sherrod

Observation Methodology

The search for external supernovae is rewarding. Not only can an individual or group of individuals contribute to science, but the sudden appearance of a "new star" creates motivation for further study and provides motivation for learning and exploring further into the world of astronomy and astrophysics.

The likelihood of any one individual, who adheres to the routine specified here, finding a supernova within a relatively short time is actually quite good, provided that a routine observing plan and schedule are maintained.

Equipment Requirements

With today's modern computerized "GoTo" telescopes, acquisition of dozens of galaxies within one night is possible. The minimum requirements to pursue this study should be:

1) at least a 25cm (10-inch) aperture, computer-controlled telescope
2) a suitable planetarium computer control program; I strongly suggest **GUIDE** (Project Pluto) [11] **The Sky** (Software Bisque) [12] or **Stellarium** [13]; this program will be required to have a fairly complete listing of nearly all New General Catalog (NGC) [14] objects; most come standard with the option to load all of the NGC catalog during installation.
3) a second ancillary program which provides "blink" capability (see following)
4) CCD or CMOS (Digital Single Lens Reflex) **monochrome** camera that can be mounted in the telescope
5) capability of accurate (or guided) 45-second exposures (this saturation level along with the minimum aperture stated above will provide excellent images of all galaxies listed following, along with capability of capturing magnitudes to 19.0 via CCD.

Additionally to this, the following hardware is recommended, or likely to be added over time:

1) photometric Bessel-Cousins "V" filter [15]
2) telescope focal reducer (around f/0.6 or f0.7 is recommended) [16]
3) computer folder designated specifically as an archive of all galaxy images obtained as the observatory's "base" or comparison images
4) photometric reduction computer software (*MaximDL, MPOCanopus*) [17]

Ideal considerations would include of course an enclosed observatory with computerized operation either remotely or on-site which is quite common today. Nearly all more sophisticated private and institutional small observatories also are moving toward complete robotic computerized programmed control of telescope operation of each night's observing. This will allow for the telescope to be tasked to access each galaxy, in order of Right Ascension, based on the programmed NGC number given following, image and record the image of one galaxy and then move onto the next target in the matter of only minutes.

As sophisticated astrophotographers grow weary of simply gather photon data over and over again on the same deep sky objects, they can put their sophisticated equipment to scientific use by this incredibly simply and rewarding study.

Software Recommendations / Creating and Using Your Master Comparison File

At the *Arkansas Sky Observatories* (ASO) [18] the primary study over the past two decades has been refining orbits of Near Earth Objects, with a secondary study of cataclysmic stars. Although there have been many documents written about supernovae searches, including one on the ASO website), this is the first comprehensive listing of PRACTICAL targets in terms of possible discovery as well as suitability for modest modern facilities.

In 2016, ASO began a systematic CCD catalog for supernovae based on a list compiled by the author of nearly 300 galaxies, in right ascension order, based on the New General Catalog.

The software used for this study is recommended here, although many other very good programs are available in lieu of these recommended. The "tried-and-true" simplicity and effectiveness of the equipment and software used by ASO is the basis of this publication.

Computer telescope control is imperative, as is accurate GoTo access of each object. Once the object is acquired and centered on the CCD preview screen, an image is taken (about 55 seconds) in .FITS format [19] (this recording the exact computer time stamp, date, telescope data, and other information embedded in the photo itself) and then saved to a file folder that is labeled for the precise DATE and YEAR in which the photo was taken. All of these images are archived for future reference if needed.

Note that storage of individual search photos must NOT be within the "Master SN" file folder where each of your standard **comparison** photos reside! You will at some point very soon after taking the photo (preferably immediately or after your first hour) want to "blink" the Master comparison photo against the one just taken. Any new star will immediately pop out as it blinks on and off between the two photos.

NOTE that if your equipment ever changes, you must acquire a NEW Master photo of every object, so that the two images will always match. The match MUST be oriented exactly in terms of sky orientation (**north at top, east to left**) and exposure time – if the Master photo was 45 seconds in length, so should the others be reasonably close each night in search of supernovae.

Always acquire your Master comparison photo when the galaxy is as high in the sky as possible to eliminate the chance of atmospheric extinction limiting actual foreground stars from being recorded. It is imperative that your Master photo be as clear and sharp as possible. I recommend that the Master photos be 60 seconds in length to maximize appearance of all faint field stars near or superimposed over the galaxy itself.

NOTE that always you must orient the camera (rotating it in the connection to the telescope) with NORTH UP and EAST to the left on your imaging screen. This is the proper orientation for both reporting and publication of sky images, as well as easy of measuring using comparison charts. It is helpful to place small marks or even an index tab onto the camera assembly so that it can only be inserted into the telescope in the NORTH UP position.

Figure 4 shows the ASO Master photo (60 sec) of galaxy NGC 6946 in which a supernovae appeared in 2017. [20]

Figure 4 – Galaxy Master photo ngc6946 from Arkansas Sky Observatories – Supernova in 2017

Set up each evening's observing schedule and supernovae search in terms of Right Ascension and select the galaxies to be examined from the ASO Target Galaxy List in this publication. Using sidereal time (the time of the right ascension hour that is directly on the meridian at any point), select the R.A. that is about **two hours west** (or earlier) than the exact sidereal time that your observatory is ready to begin. If the local sidereal time LST is, say, 11:30 that means that right ascension 11h 30m is right on the meridian at that time....just about the position of the bright star *Denebola* in Leo. Thus, select from your target list galaxies that are at R.A. **9 hours** or thereabouts to minimize atmospheric extinction.

You will center your target, then use the software that controls your camera to acquire the image. If you have capture software that came with the camera from the manufacturer, that is recommended simply for simplicity. However, you must have the capability of saving in .FITS format. The image can be saved alternately later as a .jpg or .bmp file if you like, but for comparison and magnitude determinations you MUST save as a .FITS file!

Our preference at ASO has always been the SBIG *CCDops* [21], which is the basic capture program that was provided for the now-discontinued Santa Barbara Instruments Group cameras. This software will work on many other brands of CCD cameras. Another excellent capture (and blinking) program is the Software Bisque *CCDsoft* [22] which is used at ASO for blink comparison and for some photometric reductions.

When your 45-60 second image is acquired, you will want to only do some modest density and contrast adjustments to the image, to bring it into the overall saturation level and contrast/brightness of the Master file photo of the same object. Consistency is key.

With most camera control and capture software, there is an "automatic" button or function that will self-adjust the new photo. Click that first, then make minor adjustments in brightness and contrast to bring it to the level of your Master photo.

Once your new image is acquired and stored in the daily file folder, you will want to examine it for the appearance of a new star. This can be done manually by examining carefully the image of the small galaxy against the one Master photo that you have in the Master folder. This is relatively easy to do, provided that your galaxy is small and free of foreground **field stars**.

Look at the image above and imagine attempting to locate a "new star" among all of those stars that we are looking through in our own Milky Way. In some cases, it might be virtually impossible to ascertain a supernova without some help from technology.

Thus, there are excellent programs that will aid the observer in quickly recognizing a new interloper in any crowded field of stars. "Blinking programs" have been used for decades to flash two images alternately as one apparent image, quickly blinking from your Master photo to the new one for that date.

I highly recommend the quick and easy blinking program that is included in Software Bisque's CCDsoft. Using any two .FITS images that have identical camera data (pixel size, focal length of telescope, saturation, etc.) the two files are loaded into the program by "Open", each file being opened simultaneously to appear on the control screen of the program.

Once both images are displayed and you can assure that the orientation is fairly similar (small changes in camera tilt, centering of the galaxy, etc.), there is a specific ALIGN CENTROID key

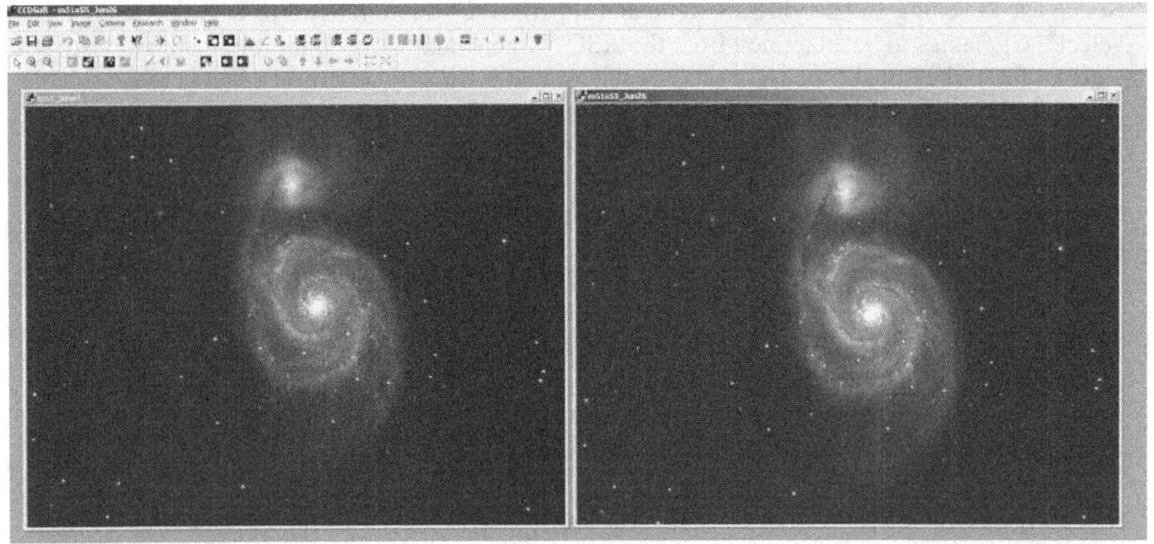

Figure 5 - Images of Messier 51, June 7 and June 26, 2011, loaded onto CCDsoft for blinking

———

Once the images are loaded, the Master image on left and daily image on right, the measurer must select a STAR in the foreground that is quite distinct for alignment purposes. This is the ALIGN CENTROID point which the program uses to perfect orient both images so that the alternate flashing appears as ONE image, with anything new blinking in and out.

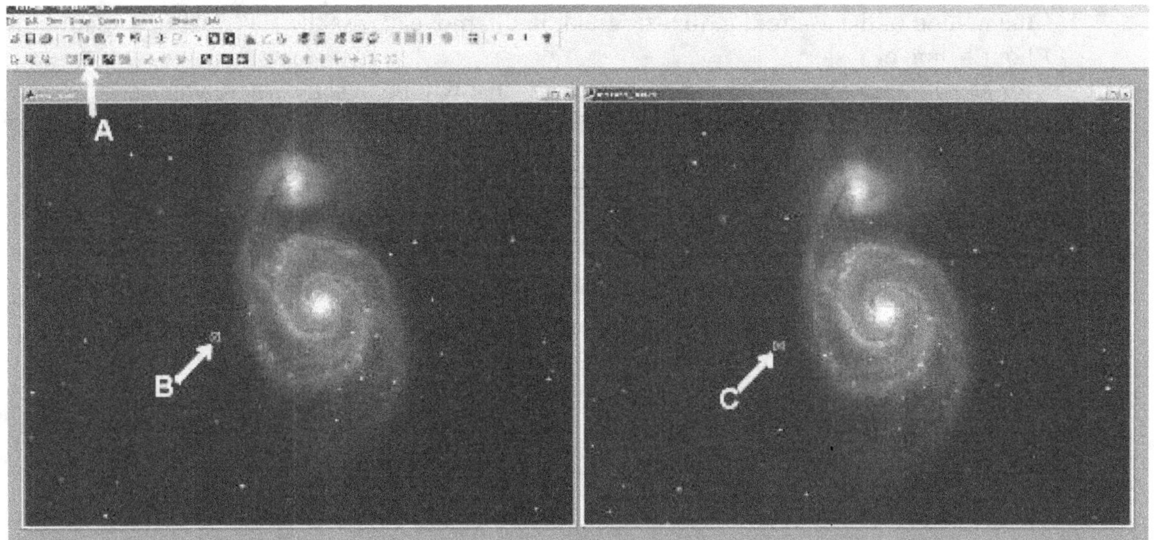

Figure 6 - Locking onto the CENTROID ALIGN prior to blinking; note the small star that was selected and locked in from the previous image (Fig), the first “”B” from the Master file photo and then “C” from the new photo from the later night. The centroid align button on CCDsoft is marked as “A”.

———

The observer then merely needs to start the “blink” process per instructions on each individual program. Typically there is a tab or sub-tab that starts the process. On CCDsoft shown here, the blink function is found under the tab RESEARCH, then scroll down to “Comparison” and the process begins.

Any new object in the field will immediately begin to flash....scan very carefully through the extent of the galaxy, all the way to the edge and just beyond. There may be “field stars” in the image, but they will remain constant. If a star-like object flashes within the arms or blurry extent of the galaxy, then you may have discovered a supernova. But WAIT!.....

Always err on the side of caution and doubt. There are two things that might be in play here that can deceive your new “discovery.”

1) this may be a supernova that has been already found that you have simply not been notified of,
2) or it very well might be an asteroid passing slowly through the field of view.

No matter, if there is a new object spotted, always proceed as if you have made a discovery and begin documenting this new object. There are quick ways to determine if your discovery is valid via the Internet search functions.

If it IS a supernova and has been already found, then you will find this by doing a “Google” search entering the words: “supernova” and “ngcxxxx” (fill in the proper numbers for your galaxy) and the current date and year. The search will reveal any published material of any know supernova in that galaxy….it is that quick and easy.

If it is a known (or unknown) asteroid/minor planet you can verify two ways:

1) the motion of the asteroid in the star field; it will move across the star field in a couple hour's time or faster;
2) for a complete search of any known asteroids that will be exactly where you are photographing at that precise time can quickly be found by accessing the Minor Planet Center's NEO confirmation page [23] that lists known minor planets for any specific right ascension and declination….just enter the coordinates of the galaxy that you are imaging: *https://www.minorplanetcenter.net/cgi-bin/checkneo.cgi*

Interestingly, if you detect MOTION of the object in your blinking and the object is NOT listed on the known objects results from the Minor Planet Center, then you should measure the position(s) and motion of the object and report this to the Minor Planet Center (MPC) as instructed at: https://www.minorplanetcenter.net/iau/info/TechInfo.html

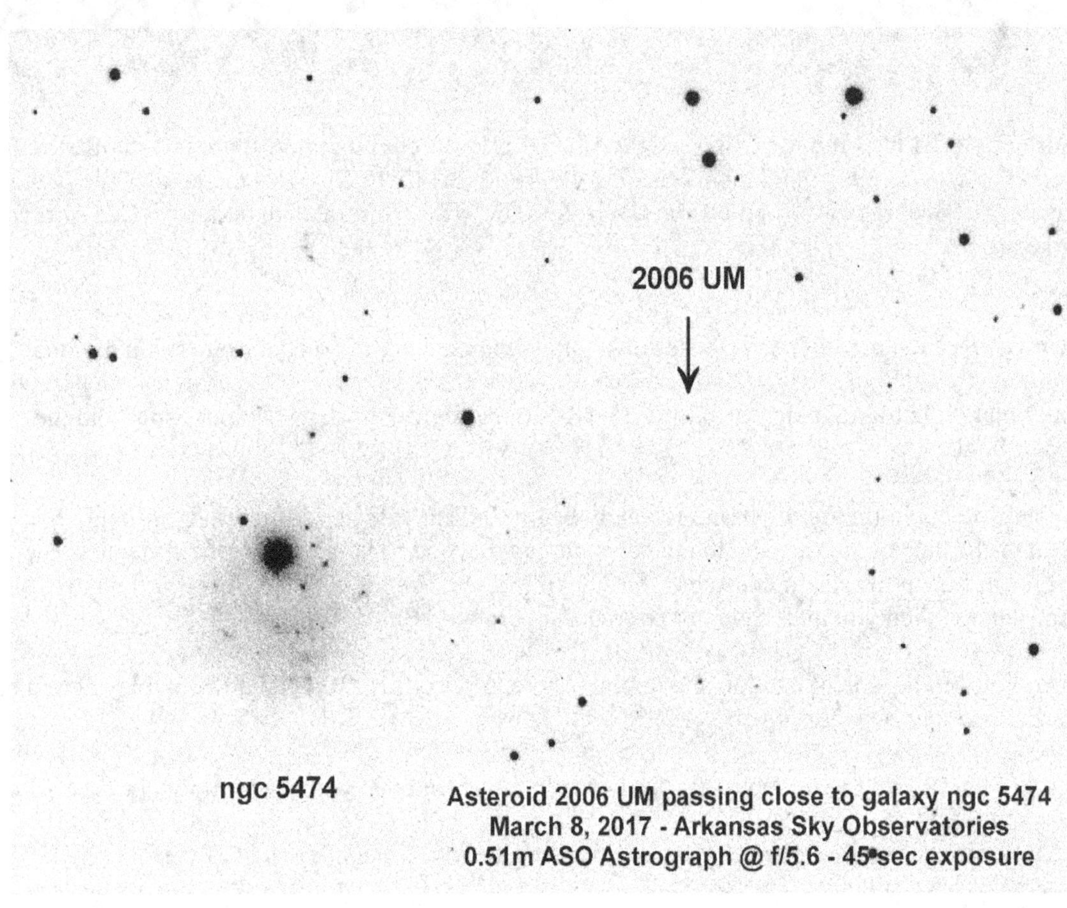

Asteroid 2006 UM passing close to galaxy ngc 5474
March 8, 2017 - Arkansas Sky Observatories
0.51m ASO Astrograph @ f/5.6 - 45 sec exposure

* * *

Comments on Photometry and Use of the "V" Filter

Photometry is the study of the luminosity and the changes in luminosity of an object; in astronomy this can apply to comets, asteroids, satellites and variable stars – among them the novae and supernovae.

In no other field is the precise brightness measurement as critical as it is in the study of supernovae. Through the process, the dynamics of the star in collapse and subsequent explosion are carefully recorded, and using the UBVRI (Ultraviolet, Blue, Visual, Red, Infrared) there is much spectral analyses of the actual star fusion processes that can be gleaned from the variation in brightness in each of those wavelengths.

The color spectrum covered by the UBVRI process was shown in Figure 3; differences in the light from a source as it is measured through each of the colors and compared is called the "Color Index" which can lead to actual spectrography of the object. For supernovae this is of utmost importance, since we are watching a star go through an incredible rapid fusion process from hydrogen, past helium, then to lithium, argon and even to lead. All of this can be revealed and studied through differential filtered photometry.

However, for this study of discovery and reporting, and subsequent follow-up of course, we are going to focus only on unfiltered CCD measurements or at best, encouraging participants to engage in filtered "V" photometry for far more accuracy and standardization of his or her measurements. The "V" filter – which concentrates only on the light from about 550nm to 675nm and very sharply only in that zone, most carefully records the light as if it were being measured accurate by the human eye.

Thus, the "V" filtration is at best the very least that is necessary for accurate and acceptable brightness reporting of your new supernova. Since some camera sensors are red sensitive and others are blue sensitive, the filter allows the passage of only the 550-675nm part of the visible spectrum and thus standardizes the magnitude reports to the AAVSO and others.

Figure 7 (following page) shows the light curve from a quality "V" filter, from the Custom Scientific Bessell-Johnson/Cousins UBVRI filter set. These filters can be obtained unmounted, mounted for 1.25 or 2.0 inch screw in applications, and in an array of sizes, both round and square.

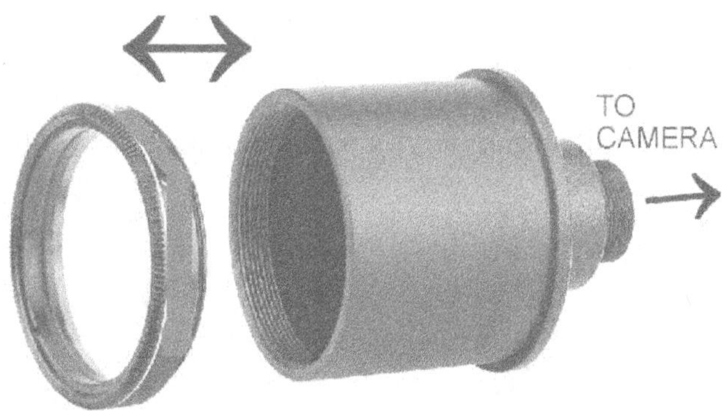

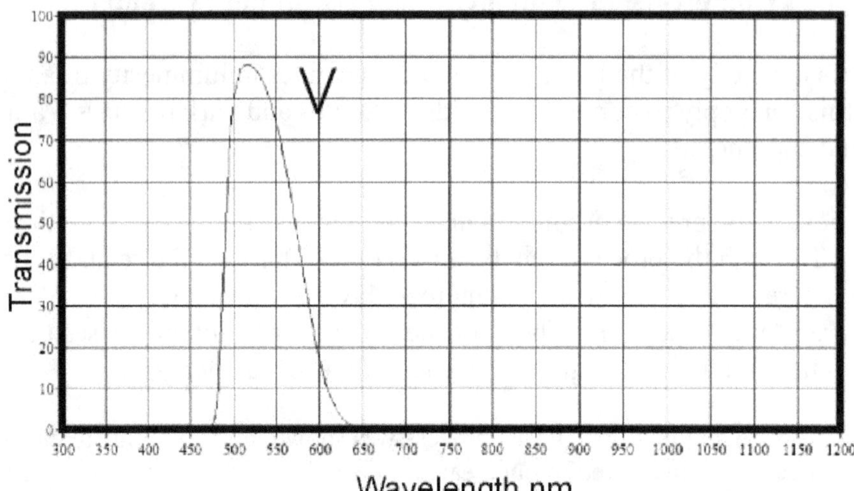

Figure 7 – the spectral limitations and isolation of the "V" filter for supernovae photometry. Adapted courtesy Custom Scientific [15]

As is so evidently displayed here is the omission of all other zones of the visible spectrum while using the "V" filter. This accepted standard allows consistency between CCD measurements from all observers, no matter what equipment or camera is being used.

Note that when using such filters, there is a substantial drop-off in total light transmitted compared to clear, unfiltered images. With this particular filter the drop-off, or light loss, is about 25 percent, one-forth of what an unfiltered image would pass. This will mean that exposures will need to be increased by a factor of 1.5 to 2 to compensate and result in the same limiting magnitude one might reach unfiltered.

Nonetheless, this is the necessary ingredient in standardization of observations. If your discovery leads to follow up observations, then photometric measurements must be made using one of these filters. Be aware of cheap imitations on the market. Examine light curves from product suppliers and see if they match up with the narrow passband seen above. Ask for guarantees of transmission performance. If you buy a cheap filter, you are going to get a cheap filter.

For all measurements, the filter must be attached at the telescope end of the camera nose-piece or on the telescope end of any reducer that is being used. The filter should not be mount directly over the sensor or any other location in the optical train other than at the entrance to the camera adapter or reducer for maximum performance.

Note that there is a slight but significant focus change necessary when adding any filter into the imaging system of any telescope. If the filter is left in place this adjustment is necessary only once.

* * *

Setting up the Master Galaxy Reference and Galaxies Folders

Some previous discussion of the necessity to establish a Master File of reference galaxies has already been made. This Master File is the key to success in an efficient and successful search for extragalactic supernovae. Organization in this study is everything.

Precise record-keeping is key to success and following are some guidelines to assure precision in your studies, both present and future.

Date, Time and Information for each patrol photo

The main advantage of recording all images in your supernovae search in the .FITS format is that all pertinent information regarding the exact circumstances under which the photo was take is recorded and embedded into each image. Thus, if the telescope information (focal length, diameter, type) is recorded in the "Information" (CCDsoft) or the "Misc" (CCDops) tabs, this data is retained so long as you are using the same instrument [NOTE: all camera operation software have similar input for telescope and camera data].

Similar circumstances for the camera are typically automatically defaulted by the software being used when the connection between PC and camera is made and defaulted. You will need to specify the binning (2x is recommended), and pixel size is the default does not indicate those settings if you are using a focal reducer or compressor and if binning is used.

Time on the other hand is completely computer-dependent. Always check at the beginning of each session and attempt to set the computer TIME to the nearest second. This time will allow the camera capture software to embed the exact time stamp at either the beginning, middle or end of each exposure (user sets preference…..I recommend beginning).

The computer that is being used for image capture must be set to **Universal Time** (**GMT**, or Time Offset = "0"); this will appear throughout the computer as the correct time and will take some getting used to, but is essential for consistent data-keeping. Clicking on the computer clock displayed in the task bar will typically open up the clock utility where you can easily change the time zone to GMT.

Some computers keep better time than others, and it is a good practice to check and reset to the precise time with the beginning of each night's observations. This can be done with the Clock Utility via direct National Bureau of Standards time sync via Internet, or can be done manually if you have a precise timepiece that has previously been set or synchronized to GPS or internet time.

Thus, all pertinent information that is needed: Telescope name, type, focal length, camera type, pixel size, binning, length of exposure, date, time and all other data will be embedded in each photograph that you take. All that is needed to access that information is to open the file with an appropriate capture program and click on the "Information" or "View" tabs.

As previously mentioned, exposures of between 45 and 60 seconds with most telescopes operating at about f/7 to f/5 will be adequate; once your exposure has been made and you are satisfied with the image displayed on the screen (reject if tracking errors, atmospheric blur, etc.), you will want to SAVE this image in its raw (contrast and brightness adjusted) format as your

MASTER IMAGE for that galaxy. Thus, this image must be a **very good one** – it is what all others will be compared to in the future.

Setting up a SN "Master" Reference File

Under your primary drive on your computer ("C" drive for most) you will want to right click and the scroll to NEW, click, and Create Folder. Once the new folder is created under your C drive, again right click on that folder and the window will display "Rename" in the list; choose that option and then type in "**MasterSN_images**". This will slowly become filled with images of nearly 300 galaxies from which all of your studies will be compared.

The photo that you just took and recorded in its adjusted, raw .FITS format will be the benchmark to which all other photographs of the SAME galaxy will be compared to (blinked). This should be a very sharp and clear image of the galaxy, oriented with NORTH up, and East to the left.

Let us say that this image was for the large edge-on galaxy ngc891, high in the sky in late fall and early winter. After a couple of shots, you choose the best one and that will serve as your master shot for ngc891. SAVE that file in .FITS format as "**891M**" in the new folder **MasterSN_images** that you just created. This will place your photo in NGC numerical order within the folder as others are subsequently entered. Placing the "**M**" (for *Master Photo*) at the end of the name will allow the numerical listing to be more easily accessed.

Since all target galaxies in this study are NGC objects and designated by that number (some as noted are also Messier objects), it is not necessary to clutter up the file name with "ngc" or other labels.

If, however, you are like me and really precise about record-keeping – even though this information is stored in the .FITS data file for each photo – you might want to add the DATE on which the photo was taken at the end of the file name: "891M_nov20_2017" for example. Since this photo is to be used exclusively for comparison purposes, the exact date really does not matter unless some transient phenomena might have coincidentally taken place in the field of view the exact moment that master photo was taken.

Thus, we now have ONE Master Folder in your "C" drive: "**MasterSN_Images**" in which all comparison images will be stored, this growing with smaller FILES of .FITS images up to around 300 (if you chose to do the entire list, or even 150 if you do one-half of it), each named for a specific NGC galaxy that you are surveying for supernovae.

We also have ONE File in the Master Folder: "**891M_nov20_2017**" (or just "891M" if you choose), this being a photograph of galaxy ngc891 that serves as your comparison galaxy for blinking search.

The list will grow rapidly as you both study for supernovae AND add comparison galaxies from the NGC list. Each night out, you can record a new image, for example of ngc891, for comparison, while easily obtaining photo comparison images of galaxies that you are adding to your search. Ideally, the list would grow to be one folder containing .FITS images of around 300 different galaxies.

Your comparisons can be done night or day, rain or clear; you need to merely open both your NEW image from the previous night of, say ngc891, then the comparison image from this folder into your blink program, and prepare to study and possibly discover.

You now must set up a SECOND folder which will contain up to 300 FOLDERS in which to store each night's capture of every galaxy on your list – the Galaxies folder.

Setting up a Working "Galaxy" Folder

Again, in your "C" drive you must now insert a NEW FOLDER, this one we might call "**Galaxy**" in which all of your nightly images for every galaxy studies will be stored, studied and subsequently archived for possible future use and reference.

The first example image that we obtained (previous page, "891M") is NOT a search photo; rather it is the photo upon which all searches will be compared. Let us say that you obtained "891M" on the evening of November 20, 2017 and were pleased with the clarity, depth and contrast and selected it for your MASTER photo for ngc891. That is in your Master file.

Tonight, November 24, you are not only going to add new Master .FITS images for other galaxies, but you are also going back to ngc891 and are going to take a new photograph of it, this being the first **Search** photo for that object, hopefully the first of dozens of nights into the future. In your Master folder you already have the standard comparison photo for that object, and thus it can be blinked.

This new – November 24 - .FITS image, taken under with the same equipment, orientation and exposure time, can be acquired in less than five minutes total time and it will be stored in the NEW folder, "**Galaxies**". Under the Galaxies folder, are up to 300 sub-folders, each one for all of the NGC objects on the following list (or however many you begin to add as your research continues). This November 24 image of ngc891 you might label "**112417_891**", this showing the date (11-12-17) and the NGC number. Actually the NGC number is really not necessary, but recommended, since the .FITS photo from November 24 is going into a sub-folder that you create in which you will re-name "**891**".

In theory, if you were to attempt to study all of the objects in the list following, you would have about 300 sub-folders starting with "**128**" proceeding in numerical order and finishing with "**7816**." For future study, you open each of those folders and simply find the image for the exact date, as shown with the file name. Under each folder are ONLY images of that specific NGC galaxy, nothing else except maybe for personal notes pertaining to that object.

To assist in backing up your MASTER comparison photography, it is a good idea to also put a copy of that master file in this "Galaxy" file, as well as back up the entire MASTER folder to a flash drive or DVD.

In time, both the Master file and the Galaxy files will grow rapidly. Never delete permanently any image – always hold onto at least one image of each galaxy for every night you put in time on that object.

Your study of each galaxy, and hopefully subsequent discovery over time of a supernova from one, will now take place quite easily: Open the blink comparison program on the selected software of your choice; then open the MASTER image of a particular galaxy; open in the same program the new photo just taken; do a Centroid Align, and begin blinking.

REMEMBER: your success in this study requires consistency: SAME camera using the same binning each time. ***SAME*** telescope using the same reducer and focal ratio each time. ***SAME*** length of exposure for every subsequent image of each galaxy. And *never, ever* put off the blinking process any longer than possible....**discovery** requires immediate attention and reporting. There are a lot of huge sky survey telescopes out there every night scanning the entire sky! Your key to success is timing and no delay from red tape. SEE **Page 26** for ***Reporting***.

Desktop Icons: Making Acquisition and Process Easy

A final word about your two computer folders for this study, **Master** and **Galaxies**: to assist in the efficiency of your process of acquiring the image at the camera, filing it, then comparing it to the Master .FITS file via your blink program, will increase dramatically if you make things easy on yourself by putting FOUR Desktop Icons on your PC desktop for easy access.

You will want to create Icons for each:

1) Camera control; 2) Master File; 3) Galaxy file; and, 4) blinking program.

Create the Icons by simply right-clicking on the folders in the "C" drive which opens up an Option menu; under that you will see "Send to" which opens a series of choices. Click on "Desktop" and that will send an Icon to your desktop. Arrange all four together in some suitable order so that they can be seen and manipulated even when any or all of these tasks/folders are open and perhaps minimized on your desktop.

The one task that you want fully displayed will be the BLINK program so that you can fully utilize your visual acuity. I have found that using the blink programs in a darkened room great increases the acuity of the eye and in addition reduces the stress and fatigue associated with such associated close work on the computer.

* * *

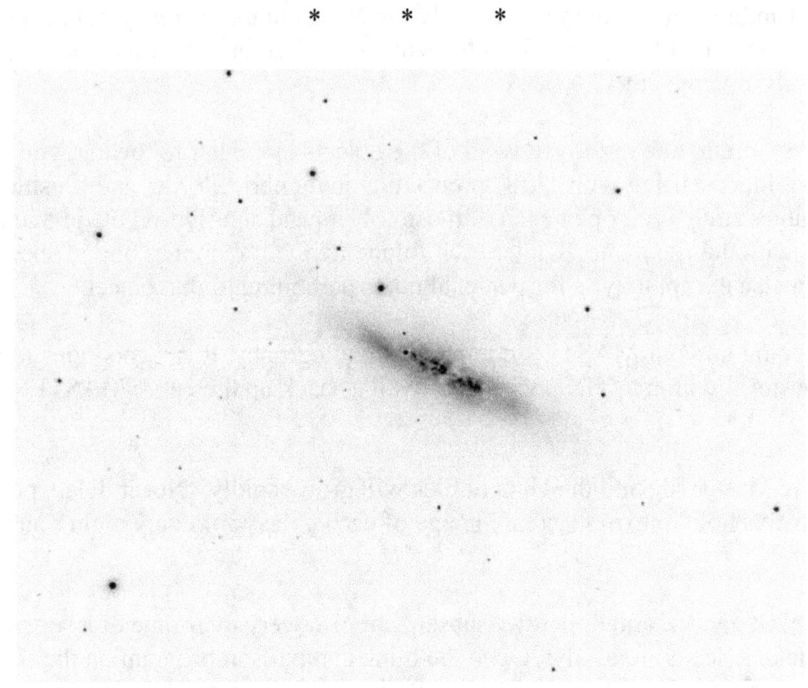

Supernova in irregular galaxy M82, January 25, 2015
Photo by Arkansas Sky Observatories, P. Clay Sherrod
0.51m f/5.9 ASO Astrograph, 45-seconds

Catalog of Galaxies Suitable for Supernovae Surveys

Extracted from the Revised Complete New General Catalogue, 2009

Suggested Top 300 Spiral and Irregular Galaxies for Modest Telescopes and CCD

Introduction

When Arkansas Sky Observatories added supernovae surveys to its list of early detection research projects, there were many guides available on how to take a photo of a galaxy with a modest telescope equipped with CCD camera and computers and how to find any new interloper that might be in the area covered by the distant galaxy.

Just picking random targets, one after another was not methodical and led to discrepancies in time periods covering any number of galaxies. One galaxy does not take but a few minutes to locate, frame, snap a photo, save and then compare. Then an observer can move onto the next galaxy.

But there was no documentation of an authoritative and concise, easy-to-follow guide of galaxies outside of the Messier list and a few bright galaxy catalogs. None of these contained practical information as to what an observer might expect when the photograph of a single galaxy would appear on the computer screen. Were all the stars superimposed across the galaxy's boundaries merely field stars or could some be true supernovae in the distant galaxy?

Which galaxies out of the 8,000 object New General Catalog (NGC) were suitable for surveys using fast acquisition and relatively bright targets? How many galaxies were suitable for our equipment? What targets might be up at any night of the year?

The following list has been compiled from decades of galaxy survey photography, and reveals the top 300 galaxies in terms of suitability for supernovae survey study. Considerations for inclusion were:

1) brightness of the galaxy (no fainter than 15th magnitude)
2) size of the galaxy – in most cases size is in proportion to distance of galaxies and small size indicates great distance and therefore supernovae too faint to be recorded
3) type of galaxy – only spiral and irregular are included as discussed earlier
4) orientation of galaxy – preference given to those in favorable presentation angle
5) field of view – small galaxies in dense star fields were not included; preference is given to galaxies that we had recorded showing few stars to interfere with interpretation.

In the following list, we have each object in Right Ascension Order (R.A.) beginning with 00h, very near the location of the Great Andromeda Galaxy by coincidence. Starting each line is the **NGC** number, followed by the **Right Ascension**, **Declination**, **MAG** (magnitude V) and some very brief and abbreviated **Descriptions** of what might be expected in this field.

Some abbreviations may include: "*clear field*" – no stars in front of galaxy / "*star to S*" – star to south / "*embedded stars*" – stars in disk of galaxy that can be confused for supernovae / "*irreg*" – irregular galaxy / "barred" – barred spiral / "*mottling*" – irregular lumping within galaxy / "*plus ngc….*" – another galaxy in same field of view / "*diffuse*" – smooth and disk like in photo / "good (exc., great) target" – these should be given priority / "*crowded field*" – many foreground stars.

Some Practical Suggestions

There are some galaxies that are obviously going to be easier for you to monitor than others: some are larger and brighter, so devoid of field stars and some more suitable for your observatory location. There also are some basic guides that I might suggest to assist in the efficiency of your observations in term of target selections for any given period during which you are surveying the distant galaxies in search for these "new stars."

Season: this list is presented in ascending R.A. order, starting a 00h. Because of the nearly three-minute difference in local vs. sidereal times, our seasons change and not all objects are visible year-round. Exceptions would be those objects which are **circumpolar**, or within 30 degrees of either the north or south celestial poles. Nonetheless, nearly all objects in the following study group are seasonal and your observing can be spread out over the course of the year. Examine the list and determine the approximate **Sidereal Time** at one hour after sunset, the time at which the sky attains darkness suitable for imaging. That Sidereal Time is the Right Ascension of the meridian at that moment overhead....all objects west of that time will have R.A. values less, and those east will have greater R.A. values. For your night's observing plan – or if you choose to start at midnight, etc. – select a starting point about two hours west of that Sidereal time, and extend to about two hours east. Thus, if your Sidereal Time in the example here of January 11 at 7 p.m. local time is 02:13, then objects at about **R.A. 02h** are directly on the meridian and slowly moving westward at that time. Your starting R.A. for galaxies might then be 00h R.A. extending eastward to 04h R.A., two hours west and east of the meridian. Remembering this Sidereal relationship will allow you to PLAN each evening's observing schedule to maximize the number of targets throughout your evening or morning.

Filtering Criteria: essential to the success of any study is the feeling of the student that success is being made even if discoveries are elusive. Obtaining quality images that are acceptable for your Master image as well as subsequent target images for blinking of the same object is paramount in encouraging dedication to your project. I strongly suggest that each observer start with targets that are easy and accessible, and those which are more easily blinked than others. This requires looking over the data following in the "Description" quips to look for the "good target", "open field", etc. to find objects which you might ascertain would be the easiest to start with. The more difficult targets can be added in time.

Declination: part of filtering criteria is declination (**DEC**) of the object being targeted....is the declination too far south for steady imaging or is it prone to atmospheric distortion? By no means do you want to restrict your targets away from targets close to the horizons, but those consistently too far south, for example, might not be the best use of your time. To test your limitations, choose a night in which stars far to the south (or north in the southern hemisphere) do not "twinkle" excessively. On such nights take some test images of galaxies at 10-degree decreasing increments and examine. From your location, you may find that there is a point at which imaging any further closer to the horizon results in unusable images for this study. As an example, at the location of ASO here in the forests of Arkansas, we have rapidly deteriorating southern steadiness due to the local geography, and targets less than –15 degrees declination are unsuitable for photometric study. Use the **DEC** value provided in this list to determine the range of objects suitable from your location.

Magnitude: equally important to success is that of **magnitude (MAG)**; remember that for galaxies, the magnitude from all published sources is given as an expression of **total integrated magnitude**, or the brightness of that object if ALL the light were collected and compacted into a single point source, like a star. Thus, a 12th magnitude <u>star</u> is going to be far easier to photograph than a 12th magnitude galaxy that is extended over an area maybe four square arc-minutes of space. That being said, modest telescopes (0.25m) operating at f/7 or faster with good CCD equipment can easily attain quality images of nearly all of the following selected objects with exposures of 45- to 60-seconds. Stars in those galaxies down to magnitude 18 or 19 are possible with quality seeing conditions. It is suggested that observers for this supernovae study begin with the brighter and larger galaxies and start their Master files and immediate searching subsequently with those objects. This selection based on brightness ALSO has the added advantage to serve as a teaching tool for the Centroid Align and blinking process.

Embedded Stars: the issue of embedded stars is an important one. Galaxies which are marked as "embedded" have stars - some of them faint but nonetheless likely to at some point show up on your patrol images – which are foreground (in our galaxy and not part of the other galaxy) objects but they appear to actually be within the area of the galaxy being studied. "**Field stars**" are stars that are located nearby but outside the realm of that galaxy's expanse, yet still might be immediately misinterpreted as a possible supernova. Blinking between several nights will quickly reveal stars that are steady and consistent between images….those are to be ignored. An embedded or field star will remain steady….a supernova will flash in and out between images when being blink compared.

Mottling: this is another important aspect of observing galaxies for transient phenomena. "Mottling" describes a galaxy in which there are lumps of galactic material, some possibly clusters of stars in that galaxy, perhaps a smaller galaxy superimposed, or some irregular activity within the galaxy; most mottling does NOT appear star-like, but somewhat as a fuzzy tiny spot within the galaxy; again blinking between subsequent nights will reveal a steady image for mottled structure. This should be ignored.

Image Orientation: A final word about image orientation: although this has been discussed before, it is important to conclude this study overview with the proper orientation of BOTH the Master image and the subsequent target images. Most observers may choose to remove their CCD cameras, even if observatory operated, between sessions to prevent moisture and insects from perhaps getting into the equipment. Such a practice is recommended. The was previous discussion on the importance of re-inserting the camera as close to the same position between every single night of observing by indexing between the camera nosepiece and the telescope focuser or back plate mount. This does NOT mean that complete precision is required. However it does mean that you always attempt to place the camera back as close to the original position of your MASTER image as possible, with NORTH UP and EAST to left. Remember that "close" is okay; that is the purpose of the "CENTROID ALIGN" process on the computer's blinking program. Picking out the same field star and locking in on them between any two images will result in the program adjusting for small offsets such as position angle (camera rotation) and non-centering in the field of view. So if your images are offset slightly, you are okay. Just make sure they are close enough so that the blinking program can lock them and calibrate them precisely one over the other.

NOTE: an asterisk (*) in description indicates a very good target for most telescopes

* * *

The Discovery – Reporting and Confirmation Process

For those who dedicate the energy and time required to do so, there is a good chance at discovery of at least one supernova in your searches.

As has been mentioned, there is always a chance that some "new star" discovery within a galaxy's borders may be in fact a solar system interloper – a minor planet (asteroid) that is already known or possibly one that is not known. See Page 16 for full discussion of that possibility.

When your blinking of images reveal a "flashing star" within the galaxy between your Master Image and your newly acquired Galaxy image, the reporting of such should be made as soon as possible. Please follow the steps below to assure accurate reporting:

1) Determine the POSITION for the new object, using a PC sky program, by simply placing your cursor over the star chart you used for the GoTo command to your telescope, at the precise position of the newly found star; depending on the PC program you are using, you can easily see the read-out in R.A. and DEC of the position of that star….or some programs might require you to right click at that position to reveal the coordinates. Immediately right those positions down, along with the Date, Time, and NGC number of the galaxy.
2) If possible, return to that galaxy either that night, or the following night, and obtain a second image of the same target, and blink that one as well; if the object is still there and has not moved, you may have a supernova.
3) Determine the **magnitude**, either by a very careful VISUAL examination or by the photometric program in your blinking program (CCDops, for example, has a very accurate photometric program where you calibrate easily with known stars in the field of view). You can display or access the stars of similar magnitude on your PC sky program if you have either UCAC-4 star catalog [26] or similar. A complete list of Star Catalogs and the download for a very good measuring program: *ASTROMETRICA* [27] can be found at their website: http://www.astrometrica.at/default.html?/catalogs.html . I urge your to visit this page for a free and excellent PC program.
4) **Send** via e-mail (do not Tweet, or post on Facebook or other social media) your discovery report which must include: Date, Time, Observer, Equipment used, NGC number, Magnitude.
5) Continue to **monitor** your object nightly….if a supernova, it can change rapidly.

There are many agencies to which the report of a new supernova should be sent, and you must strive to get the report of your discovery out as quickly as possible, but not at the sake of mistakes and accuracy. **Read** the entire discussion beforehand from the International Astronomical Union (IAU) about reporting discoveries: https://www.iau.org/public/themes/discoveries/

Once you are certain that you have a new object, send the information from 4) above, including a copy of the discovery PHOTO if possible to:

Transient Name Server, IAU - https://wis-tns.weizmann.ac.il/
American Association of Variable Star Observers - aavso@cfa.harvard.edu
IAU Discovery Page - http://www.cbat.eps.harvard.edu/HowToReportDiscovery.html
IAU/CBAT Direct Report via e-mail - cbatiau@eps.harvard.edu
CVS Variable Star Alert (Japan) - vsnet-alert@ooruri.kusastro.kyoto-u.ac.jp

REPORTING PROTOCOL - IAU

Extracted and adapted from the IAU Transient Objects Confirmation Page
http://www.cbat.eps.harvard.edu/unconf/tocp.html

FIRST: Make sure that you do not mistake a minor planet for a supernova; there are two places you must check to verify that there is NO minor planet where your suspected SN might be (or you can take two photos over time and look for motion of the object):

NEO CHECKER:
https://www.minorplanetcenter.net/cgi-bin/checkneocmt.cgi
MINOR PLANET (non-neo) CHECKER:
https://www.minorplanetcenter.net/cgi-bin/checkmp.cgi

Once verified, then proceed as described by the IAU Transient Objects page:

MAKE A TEMPLATE of the exact layout and spacing below as shown in the IAU TOCP page. It is important that your submissions for possible discovery match in the exact sequence for digital processing of your information. It is very important to use only spaces and do not "tab" between column entries!

There are two types of data that can be contributed as sequences of single lines: one for discovery data (with an asterisk in column 20), and another for follow-up data (no asterisk in column 20), as described below. No tabs should be used (only spaces). When sending e-mail, put "TOCP discovery observation" or "TOCP follow-up observation" in the subject heading, and in the text, give only the lines of data as indicated below (only one line per discovery object; multiple lines for follow-up observations are permitted). Examples are given below:

Discovery data:

```
123456789 123456789 123456789 123456789 123456789 123456789 123456789 123456789

Type      Date (UT)       R.A. (2000.0) Decl.     Mag. p    Offset     Locale    A C

TCP 2009 11 10.8811*  06 26 57.68 +24 29 06.8  16.2 U                Gem       1 7
PSN 2010 02 08.92   *  13 29 15.92 +46 30 14.0  15.8 U                          0 0
PSN 2010 06 18.47   *  22 54 20.33 +11 46 54.7  18.4 U   10E    2S  U12237    9 -
PNV 2010 10 13.5566*  00 44 26.56 +41 31 13.6  17.8 C 1350E 306N   M31       3 1
```

NOTES: The columns of the discovery record must be strictly followed, with **no tabs (only spaces between data entries).**
Columns 1-3 contain one sets of three upper-case to denote type of object suspected: PSN for possible/ probable/definite supernova; PNV for possible/probable/definite nova; and TCP for all other non-moving variables.

The date must have the year in columns 5-8, two-digit month in columns 10-11, and date in columns 13-19 (with decimal point in column 15).

Put an asterisk (*) in column 20 to indicate a discovery.

The R.A. data go in columns 23-33 (decimal point for seconds in column 31); the Decl. data go in columns 35-45 (a plus or minus sign must be placed in column 35; decimal point in column 44).

The magnitude is given in columns 48-51 with a decimal point in column 50; column 53 contains a letter to denote bandpass [U = unfiltered CCD; v = visual; V = Johnson V; R = R-band; p = photographic (Tri-X/Techpan band assumed); etc.].

Columns 55-64 are for an offset in arcsec (no decimal fractions; only whole digits) measured from the center of an extragalactic galaxy; they **should be right-justified** with the R.A. offset given first in columns 55-58 (the column 59 reserved for E = east or W = west of presumed-host-galaxy's center), and the Decl. offset given in columns 60-63 (the column 64 reserved for N = north or S = south).

Columns 67-75 are for the "Locale", meaning the constellation's three-letter IAU abbreviation if in the Milky Way, or the catalogue designation of a presumed host galaxy if known (first letter M = Messier, N = NGC, I = IC, U = UGC, P = PGC, otherwise leave blank).

Column 77 is for the number of confirmed discoveries previously reported by this contributor (single digit from 0 to 9, where 9 is nine or more). Column 78 is for the arc of observation: "-" = one image only; 0 = multiple images on a single night; 1-8 = positive images available of new object on one to eight different nights; 9 = images available on 9 or more nights.

* * *

Below: SAMPLE suggested wording for brief report to *Central Bureau of Astronomical Telegrams* and the *Transient Name Server* forms: (note that you will, of course, change the wording that is included within the **brackets "[** *italics* **]"** in the sample below to reflect your actual observatory, location and specifics of discovery) ALL of this data should be in your discovery report.

[*P. Clay Sherrod, Arkansas Sky Observatories*] reports his discovery of a possible supernova (mag about [*XX.xx*]) on an unfiltered CCD frame taken [*2018 Feb 11 06:25:31*] UT using a [*0.51-m f/5.9*] astrograph near [Petit Jean Mountain] , USA. The new object is located at R.A. [*10h15m039.90s, Decl. +73°26'08.83*] (equinox 2000.0), about [*338" NW*] of the center of NGC [*3147*]. [*Sherrod*] notes that nothing was visible at this position on an image taken on [*Jan 08*]

ASO CCD Catalog: 300 NGC Galaxy Supernovae Candidates

NGC	RA	DEC	MAG	Description/Notes
RA 00 hours				
7814	00 03	+16 08	11.6	some faint stars in field
7817	00 04	+20 45	12.5	embedded stars to S,NE
7816	00 04	+07 28	13.7	very small, clear field
128	00 29	+02 54	12.8	plus ngc 124, 126,127,130,1760
182	00 38	+02 44	12.3	edge-on, clear field, small
160	00 36	+23 57	12.6	very small, clear field
169	00 37	+23 59	14.4	very small, faint
157	00 35	-08 24	11.0	* nice barred, stars to NE
150	00 34	-27 48	12.0	tiny, star to S
171	00 37	-19 57	12.9	
182	00 38	+02 34	13.4	face-on, clear field
214	00 41	+25 30	13.0	small, clear field
210	00 41	-13 54	12.9	very clear field, stars NE,W
224	00 43	+41 16	04.4	* M31-concentrate in sections
255	00 48	-11.28	12.4	* small, but clear, star to S
247	00 47	-20 46	09.6	* edge-on/very large
253	00 48	-25 17	08.9	* edge-on/large/mottled many stars
271	00 55	-00 55	13.0	
309	00 57	-09 55	12.5	* open faint face-on, clear field
329	00 58	-05 06	14.3	plus ngc 321/327
327	00 59	-07 35	12.1	* good target
RA 01 hours				
357	01 03	-06 21	13.1	plus ngc 355
352	01 02	-04 15	13.5	
410	01 11	+33 09	12.5	plus ngc 409/414
450	01 16	-00 52	12.3	
474	01 20	+03 25	12.3	plus ngc 470
493	01 22	+00 57	12.9	
497	01 22	-00 53	13.8	
514	01 24	+12 55	12.3	* nice target
507	01 24	+33 15	12.3	plus ngc 504/508/503/494
521	01 25	+01 43	12.5	
520	01 25	+03 58	12.2	NOTE: colliding/irrg. pair
522	01 25	+09 59	14.0	faint edge-on
532	01 25	+09 16	13.9	
578	01 31	-22 40	11.5	* far south; good target
598	01 33	+30 39	06.2	M33 - dense star field
615	01 35	-07 21	12.5	
628	01 37	+15 47	09.8	* M74, very large many "suspects"
660	01 43	+13 38	11.7	extending arms, large, faint
669	01 47	+35 34	13.3	small, faint
672	01 48	+27 26	11.4	plus IC1127

673	01 48	+11 31	13.2	* small but good target
676	01 49	+05 54	10.4	some embedded stars
684	01 50	+27 38	13.3	edge-on
693	01 51	+06 08	13.2	very small
691	01 51	+21 45	12.3	some embedded stars
697	01 51	+22 21	12.8	* good target
718	01 53	+04 12	12.6	
753	01 58	+35 55	13.0	faint embedded stars
779	01 59	-05 58	11.9	near edge-on
772	01 59	+19 00	11.1	large, nice structure

RA 02 hours

784	02 01	+28 50	12.1	very faint,N-S
803	02 02	+16 02	13.2	
818	02 09	+38 46	13.2	
864	02 15	+06 00	11.7	nearby stars
877	02 18	+14 32	12.6	plus ngc 876
895	02 22	-05 31	12.2	two nice arms
891	02 22	+42 21	10.9	* fantastic edge-on
908	02 23	-21 14	10.9	very nice pecul.
925	02 27	+03 35	10.6	* very nice large, embedded stars
926	02 26	-00 20	14.0	interesting, small
936	02 28	-01 10	11.1	barred spiral
958	02 31	-02 57	12.9	tight spiral
972	02 34	+29 18	12.1	
1024	02 39	+10 51	13.5	plus ngc 1028/1029
1003	02 39	+40 52	12.0	some embedded stars
1032	02 39	+01 05	12.7	star to NE
1035	02 39	-08 08	12.9	star to SE
1023	02 40	+39 04	10.5	dense; ! stars in field !
1042	02 40	-08 26	11.5	* large, very open
1055	02 42	-00 27	11.4	very nice near edge-on
1068	02 43	-00 01	10.0	M77, dense core
1073	02 44	+01 20	11.6	very open, diffuse
1087	02 46	-00 30	11.5	* face-on spiral
1090	02 27	-00 15	12.6	

RA 03 hours

1201	03 04	-26 04	11.7	good southern target
1232	03 10	-20 35	10.5	great target for southern
1241	03 11	-08 55	12.7	also ngc1242, small, faint star NE
1247	03 12	-10 29	13.0	edge-on, star N of core
1300	03 19	-19 25	11.1	nice barred spiral
1337	03 28	-08 23	12.6	edge-on, faint star to SE
1417	03 42	-04 42	12.9	faint
1421	03 42	-13 29	12.0	edge-on, irreg.

RA 04 hours

1507	04 04	-02 11	12.9	*edge-on, clear field
1530	04 23	+75 18	12.4	embedded stars
1560	04 33	+71 53	12.0	* large, faint, nice target
1569	04 31	+64 51	12.0	* tilted, nice target
1589	04 30	+00 52	13.0	* nice, stars to NE, W
1622	04 38	-03 13	13.2	three: ngc 1618/1625
1507	04 40	-02 11	12.9	edge-on, watch for stars N,S
1620	04 36	-00 09	13.6	tilted, star to NE
1637	04 41	-02 21	11.5	* good target

RA 05 hours

1784	05 05	-11 52	12.4	* many stars in field
1832	05 12	-15 41	12.1	dense core, stars to E, SE
1888	05 23	-11 30	12.9	star to NE of core
1961	05 42	+69 23	11.9	* some embedded stars

RA 06 hours

2146	06 19	+78 22	11.3	* nice, tilted, stars to SE
2223	06 25	-22 50	12.2	star N of core/faint S of core
2273	06 50	+60 51	11.9	excel. target, face on

RA 07 hours

2337	07 10	+44 27	12.7	star to NW
2357	07 18	+23 21	13.9	edge-on, narrow, faint
2336	07 27	+80 10	11.2	faint stars, S and E
2276	07 27	+85 45	12.0	stars to SE /also ngc2300
2403	07 37	+65 36	08.9	very large, many embedded stars
2424	07 41	+39 14	13.6	very faint, sm., edge-on
2460	07 57	+60 15	12.6	dense core

RA 08 hours

2500	08 02	+50 44	12.2	* face on, clear field
2525	08 06	-11 06	12.3	barred spiral
2532	08 10	+33 57	12.4	faint, diffuse
2535	08 11	+25 12	13.1	faint star to SW
2543	08 13	+36 15	12.7	star to W
2523	08 14	+73 33	12.6	* ideal barred spiral
2541	08 15	+49 04	12.2	* large, open face-on
2549	08 20	+57 42	12.0	oriented N-S
2608	08 35	+28 32	12.8	* clear field, round
2613	08 33	-22 58	11.2	crowded star field, edge
2591	08 37	+78 02	12.9	faint edge-on
2648	08 44	+14 16	12.6	clear field
2633	08 48	+74 06	12.8	barred, face-on

2654	08 49	+60 12	12.8	nice target/stars to W
2683	08 53	+33 25	10.6	* great/stars to NE
2681	08 54	+51 17	11.1	* face on, open, star to SE
2655	08 57	+78 15	11.0	* clear field, face on
2701	08 59	+53 46	12.8	* face-on, small
2712	08 58	+44 55	12.7	* face-on, uniform, clear field
2713	08 57	+02 55	12.6	

RA 09 hours

2742	09 05	+60 29	12.2	open field
2715	09 08	+78 05	11.8	* great target
2770	09 10	+33 07	12.1	* edge-on, some stars
2776	09 11	+44 57	12.2	* good target, star to SE
2784	09 12	-24 10	11.2	* good southern target
2748	09 14	+76 29	12.4	* clear target
2841	09 22	+50 48	10.2	* large, good target, some stars
2859	09 25	+34 31	11.6	* large, open face-on
2782	09 14	+40 07	12.4	tight, face-on
2903	09 32	+21 30	09.6	very large; watch for stars
2967	09 42	+00 20	12.2	* open, face-on
2976	09 47	+67 55	10.8	* two stars to NW, nice target
2985	09 50	+72 16	11.2	face on, star to E
3003	09 49	+33 25	12.2	* edge-on, irregular
3044	09 53	+01 34	12.6	long "needle"
3031	09 55	+69 04	07.8	M81 - great target, 2 stars to S
3034	09 56	+69 41	09.2	M82 - irreg, field starts

RA 10 hours

3079	10 02	+55 41	11.5	nice edge-on, stars to N, NW
3115	10 05	-07 43	10.1	* large and exc. target, star S
3169	10 14	+03 28	11.2	* two exc. targets (ngc 3166)
3147	10 16	+73 24	11.4	face on, clear field
3190+	10 18	+21 50	12.0	3 gal, one field, ngc 3193, 3187
3184	10 18	+41 25	10.3	open spiral, star to N
3198	10 20	+45 33	10.9	* good target, clear field
3227	10 23	+19 52	12.3	plus ngc 3226, clear field
3254	10 29	+29 30	12.2	* good target; faint star SW
3294	10 37	+37 20	11.9	good target, some mottling
3338	10 42	+13 42	11.3	* very clear field, good target
3344	10 44	+24 55	10.5	* face-on, some stars embedded
3346	10 44	+14 52	12.5	nice, vague, tight spiral
3365	10 46	+01 49	13.3	barred spiral, clear field
3351	10 44	+11 42	10.5	* M95 – large, nice barred, clear field
3359	10 46	+63 13	11.0	* excellent target
3368	10 47	+11 49	12.0	M96 / face-on large target
3371+	10 48	+12 37	10.2	plus M105/ngc 3373, clear field
3396	10 50	+32 59	12.6	* plus ngc3395
3432	10 53	+36 37	11.7	irreg. edge-on, stars to SW, E
3430	10 52	+32 56	12.2	plus ngc 3432, clear field

3448	10 54	+54 18	12.5	peculiar gal; clear field

RA 11 hours

3486	11 00	+28 59	11.0	* fine face on; stars to W, SW
3495	11 01	+03 38	12.7	dim edge-on, clear field
3507	11 03	+18 02	14.7	also ngc 3501/bright star to NE
3504	11 03	+27 59	11.8	excl. barred spiral, faint star E
3511	11 03	-23 05	11.5	* very nice target, stars to E, W
3521	11 05	-00 02	10.0	* great target, some faint stars in
3549	11 10	+53 20	12.5	small but clear target
3556	11 11	+55 40	10.6	M108-edge-on irreg; watch stars
3583	11 14	+48 18	11.7	also ngc 3577, face-on spiral
3596	11 15	+14 47	11.6	* good target, face-on, star to SE
3593	11 15	+12 49	11.7	* good edge-on target, clear field
3623	11 19	+13 05	10.2	M65-great target, star to E, SW
3627	11 20	+13 02	09.7	M66-good target, embedded stars!
3626	11 20	+18 20	11.7	dense core, good face on
3628	11 20	+13 35	10.4	* very good large target, stars in
3631	11 21	+53 11	11.0	* very good face-on, clear field
3646	11 22	+20 10	11.0	face-on, stars to NW, S
3642	11 22	+59 04	11.5	face-on, star to W of core
3666	11 24	+11 21	12.3	nearly edge-on, star to S
3672	11 25	-09 48	12.1	very nice, clear face-on target'
3675	11 26	+43 35	10.8	* excel. target, stars to S, SE
3686	11 28	+17 13	12.0	* good target, face-on, star to S
3705	11 30	+09 17	11.8	* good target, star to SW
3718	11 33	+53 04	11.6	exc. target, face-on stars to E,S
3726	11 33	+47 02	10.9	big face-on, faint, star to N
3735	11 36	+70 32	12.6	very small edge-on, clear field
3756	11 37	+54 18	12.0	* very clear face-on, good target
3780	11 39	+56 16	12.2	good target, stars to S,W
3810	11 41	+11 28	11.3	* very fine face-on, embedded stars
3877	11 46	+47 19	11.6	dense edge-on, clear field
3887	11 47	-16 51	11.3	nice face-on, open field
3893	11 49	+48 43	10.9	* good face-on, stars to S,NW
3898	11 49	+56 05	11.7	* open, clear target, face-on
3938	11 53	+44 07	10.8	very nice face-on, some stars
3953	11 54	+52 20	10.6	great target, stars NE,S,W, large
3976	11 56	+06 45	12.2	nice tilted, clear field
3992	11 58	+53 23	10.6	* M109, great target, stars to N,NE
4013	11 59	+43 57	12.3	edge-on, dense, clear target
4026	11 59	+50 56	11.7	N-S edge-on, star to E

RA 12 hours

4030	12 00	-01 06	11.4	small face-on, clear field
4036	12 01	+61 54	11.5	* tilted, nice clear target
4051	12 03	+44 32	11.0	* large and open, star to S
4062	12 04	+31 54	11.9	nice open field, star to W
4064	12 04	+18 26	12.3	good target, star to E

4088	12 05	+50 32	11.1	embedded stars, mottling, SN past
4100	12 06	+49 35	11.9	very nice tilted, clear field
4111	12 07	+43 04	11.7	nice tilted, faint star SE
4123	12 08	+02 53	11.8	* nice target, face-on, clear field
4136	12 09	+29 56	11.7	* face-on, very vague, good target
4157	12 11	+50 29	12.1	edge-on, dense core, clear target
4179	12 13	+01 17	11.8	edge-on, star NW of core
4192	12 14	+14 54	10.9	* M98 -very good, clear target
4214	12 16	+36 19	10.2	barred, irreg. mottling
4216	12 16	+13 08	11.0	* large tilted, great target, star E
4217	12 16	+47 06	11.9	good edge-on, stars to N,W
4244	12 17	+37 48	10.7	* great target, edge-on, some stars
4254	12 19	+14 25	10.4	M99-wonderful spiral, watch for stars
4258	12 19	+47 18	09.1	* M106-great target, much mottling
4274	12 20	+29 37	11.3	ringed spiral, stars E,W
4302	12 22	+14 36	12.6	plus ngc4298, both good, watch stars
4303	12 22	+04 28	10.1	* M61-great target, embedded stars!
4314	12 22	+29 54	11.4	very dense, stars to SE,NE,NW
4321	12 23	+15 49	10.1	M100-large, open, embedded stars!
4395	12 26	+33 33	10.7	* many stars, large and open
4414	12 26	+31 13	11.0	good target, clear field
4429	12 27	+11 06	11.1	tilted, no structure, clear field
4448	12 28	+28 37	12.0	ringed, clear field
4449	12 28	+44 06	09.8	* irreg. great target, star to E,N
4450	12 28	+17 05	10.9	* great target, some mottling
4490	12 31	+41 38	10.2	* plus ngc4485/ very mottled, good
4437	12 32	+00 06	11.2	very large edge-on - duplicate 4517
4501	12 32	+14 25	10.2	* M88-nice target, stars to SE
4517	12 33	+00 07	11.2	large edge-on, star to NE, far W
4527	12 34	+02 39	11.3	nice tilted, star NW of core
4535	12 34	+08 12	10.5	* great target, embedded stars!
4536	12 34	+02 11	11.1	nice target, embedded stars
4548	12 35	+14 30	11.0	* M91-barred spiral, clear field
4559	12 36	+27 58	10.3	large, great target, many stars
4565	12 36	+25 59	10.3	* perfect edge-on, clear field
4569	12 37	+13 10	10.2	* M90-large, great target
4579	12 38	+11 49	10.4	* face-on, perfect target
4593	12 40	-05 21	12.2	barred spiral, clear field
4594	12 40	-11 37	09.2	* M104- sombrero - clear field
4605	12 40	+61 36	10.8	* irreg., clear field
4631	12 42	+32 32	09.7	* irreg., star at core, to N
4654	12 44	+13 08	11.1	nice face-on, mottled to NW
4656	12 44	+32 10	10.7	large, irreg., mottled at N
4666	12 45	-00 28	11.5	* dense, small clear field
4710	12 50	+15 10	11.8	pec. edge on, clear field
4725	12 50	+25 30	10.0	* large, embedded stars
4736	12 50	+41 07	09.9	* M94-dense, open clear field
4750	12 50	+72 52	12.1	tight spiral, dense core, star N
4753	12 52	-01 12	10.9	some faint stars embedded
4762	12 53	+11 14	11.1	edge-on,star to SW, S
4826	12 57	+21 40	09.3	* M64 (black eye), great target

4845	12 58	+01 35	12.1	* small, tilted, star SE of core
4866	12 59	+14 10	12.0	small, star to WNW

RA 13 hours

4939	13 04	-10 20	11.9	* open and faint, embedded stars E,S
4941	13 04	-05 33	11.9	* faint, but good clear target
4981	13 09	-06 47	12.2	* faint, good target, star to SSE
4995	13 10	-07 50	11.9	* face-on, open field
5005	13 11	+37 03	10.0	* good target, faint stars NE,E,N
5033	13 13	+36 35	10.7	* great target, embedded stars!
5055	13 16	+42 02	09.3	* M63-much mottling, clear field
5170	13 30	-17 58	12.1	* nice edge-on, faint, stars SE
5194	13 30	+47 12	09.0	* M51(whirlpool) - many SN in past
5218	13 32	+62 46	13.1	* good pair, also ngc5216
5236	13 37	-29 52	08.0	* M83-large, open, embedded stars E,S
5247	13 38	-17 53	10.7	* good target, embedded stars
5248	13 38	+08 53	10.9	* very nice, star to N, mottled
5297	13 46	+43 52	12.2	small, stars to SE
5301	13 46	+46 06	12.9	small, faint, clear field
5308	13 47	+60 58	12.2	small, edge-on, dense, stars SW
5320	13 50	+41 22	12.7	* small, open field
5371	13 56	+40 28	11.4	* good face-on, stars S,SW,W
5317	13 56	+05 01	11.2	* face on, 2 stars to NW
5377	13 59	+47 14	12.0	past SN in galaxy
5395	13 59	+37 26	12.5	* small, irreg., clear field

RA 14 hours

5448	14 03	+49 10	12.1	small, interesting, clear field
5427	14 03	-06 02	12.0	* nice, small spiral
5457	14 03	+54 21	08.2	* M101(pinwheel)- large, embedded stars
5468	14 07	-05 27	12.1	* fine face-on spiral
5529	14 16	+36 14	12.8	* edge-on, clear target
5566	14 20	+03 56	11.5	dense core, open field, star W
5584	14 22	-00 15	12.1	* nice, face-on
5660	14 30	+49 37	12.4	* small, face-on, clear field
5678	14 32	+57 56	12.2	irreg, star to S
5676	14 33	+49 27	11.9	* very good target, stars SE,W
5746	14 45	+01 57	11.3	* edge-on, 2 stars to S
5775	14 54	+03 33	12.2	* nice, edge-on, some mottling

RA 15 hours

5861	15 09	-11 19	12.1	* good target, clear field
5866	15 06	+55 46	10.8	* nice edge-on, clear field
5899	15 15	+42 03	12.6	tilted, clear field
5905	15 15	+55 31	12.3	small barred sp., star SW,SE
5907	15 16	+56 20	11.1	* great edge-on, faint stars embedded
5908	15 17	+55 25	12.8	* edge-on, stars to SE, NW
5921	15 22	+05 04	11.5	* face-on, stars NW,SW

5965	15 34	+56 41	12.7	* edge-on, small, clear field
5970	15 39	+12 11	12.2	* faint, face-on
5985	15 40	+59 20	11.9	* good target, face-on, faint star S
6015	15 51	+62 18	11.7	* great target, face-on, star to SSW
6068	15 55	+79 00	13.2	small but nice target

RA 16 hours

6070	16 10	+00 43	12.5	* small, tilted, clear field
6181	16 32	+19 50	12.5	* small, irreg., stars to N
6217	16 33	+78 12	11.8	* face-on, good, star SE or core

RA 17 hours

6384	17 32	+07 04	11.3	* exc. face-on, open with many stars
6503	17 49	+70 09	10.9	* good target, some mottling, star NW

RA 19 hours

6814	19 43	-10 20	12.1	* small, but good, star W of core

RA 20 hours

6946	20 34	+60 09	09.7	* wonderful, but MANY embedded stars
6951	20 37	+66 06	12.2	* good target, many embedded stars

RA 21 hours

7013	21 04	+29 54	12.4	very crowded field, embedded faint stars
7137	21 48	+22 09	13.0	quite small, stars to NW,SW

RA 22 hours

7177	22 01	+17 44	12.0	* small, dense core, clear field
7171	22 01	-13 16	13.0	* small, star to NE, clear field
7184	22 03	-20 49	11.8	*exc southern target, star to E
7217	22 08	+31 22	11.0	* exc target, star to N, clear field
7218	22 10	-16 40	12.6	small, irreg., some embedded stars
7223	22 10	+41 01	13.0	very small, lots of core congestion
7241	22 16	+19 14	13.5	small, w/stars NNE,SE
7280	22 26	+16 09	13.0	face-on, vague, faint star E, open
7314	22 36	-26 03	11.6	* very nice southern target, star E
7332	22 37	+23 48	11.8	* plus ngc7339, both edge-on, clear field
7331	22 37	+34 25	10.3	* great target, large, many emb. stars
7417	22 56	-05 30	13.1	small, edge-on, star SW

RA 23 hours

7450	23 01	-12 55	14.1	very small, intricate, stars
7457	23 01	+30 09	11.9	* tilted, embedded stars
7479	23 05	+12 19	11.6	* nice barred, stars N,NW,SW

7497	23 09	+18 11	13.0	* good target, stars NE,NW
7541	23 15	+04 32	12.4	* plus ngc 7537,exc target, tilted, star to E
7549	23 15	+19 02	13.9	* curious, good target, clear field
7606	23 19	-08 29	11.5	* great target, stars to SE,N; clear field
7640	23 22	+40 51	11.8	* good, tilted, dense star field
7678	23 28	+22 25	12.5	* small but good clear target
7721	23 39	-06 31	12.3	* good clear target, stars to E,SE
7723	23 39	-12 48	11.9	* dense, good clear target
7727	23 40	-12 18	11.6	* clear field, face-on
7741	23 44	+26 05	11.7	* barred, good, some stars to S
7743	23 44	+09 56	12.3	* good, dense, star at edge S
7753	23 47	+29 29	13.2	* good target, small, stars N,S,SE
7771	23 51	+20 07	13.0	* plus ngc7769. good target, clear field, sm.
7800	23 59	+14 49	13.2	faint with strong core, stars to NE

Supernovae in Messier 101, August 7, 2011
Image by Charlie Trump, 0.35m RC, 120 seconds

Bibliography

1. RochesterAstronomy.org http://www.rochesterastronomy.org/supernova.html
2. Wikipedia. https://en.wikipedia.org/wiki/List_of_supernova_remnants
3. *Universe for Facts*. Kepler's Supernova: Recently Observed Supernova". Retrieved 21 December 2014.
4. Mitton, Simon. The Crab Nebula, Simon & Schuster, 1979
5. Schramm, David. Supernovae: The Proceedings of a Special IAU Session on Supernovae Held on September 1, 1976 in Grenoble, France
6. Wikipedia.org
7. Khokhlov, A.; Müller, E.; Höflich, P. (1993). *"Light curves of Type IA supernova models with different explosion mechanisms". Astronomy and Astrophysics.* ***270*** *(1–2): 223–248.*
8. Astronomy 122: Stellar Explosions. Univ. of Oregon
9. American Association of Variable Star Observers. https://www.aavso.org/variable-stars-main
10. M.I.T. Supernovae Photometry, http://web.mit.edu/asf/www/irlab/Supernova/PhotometryPrimer5.pdf
11. Gray, Bill. Project Pluto, *GUIDE*. https://www.projectpluto.com
12. Software Bisque, The Sky. http://www.bisque.com/sc/media
13. Stellarium. http://stellarium.org
14. Steinicke, Wolfgang, ed. NGC – Revised Complete New General Catalog. 2009
15. Custom Scientific, Phoenix, AZ
16. OPTEC Corp. http://www.optecinc.com/astronomy/telecompressors.htm
17. Warner, Brian. MPO Canopus, http://www.minorplanetobserver.com/MPOSoftware/MPOCanopus.htm
18. Sherrod, P. Clay. Arkansas Sky Observatories. www.arksky.org
19. Schmitz, Cory. *How to Use FITS Images*. https://photographingspace.com/how-to-use-fits
20. Supernova in 6946. http://www.rochesterastronomy.org/sn2017/sn2017eaw.html
21. CCDops - http://diffractionlimited.com/support/sbig-archives
22. CCDsoft – Software Bisque. http://www.bisque.com/sc/media
23. IAU Minor Planet Center, Confirmation Page. https://www.minorplanetcenter.net/cgi-bin/checkneo.cgi
24. Wikipedia.org , https://en.wikipedia.org/wiki/History_of_supernova_observation
25. Vickers, John C. Deep Space CCD Atlas: North, 1994 Duxbury, MA
26. U.S. Naval Observatory. The fourth U.S. Naval Observatory CCD Astrograph Catalog http://ad.usno.navy.mil/ucac/readme_u4v5
27. Astrometrica Star Catalogs. http://www.astrometrica.at/default.html?/catalogs.html

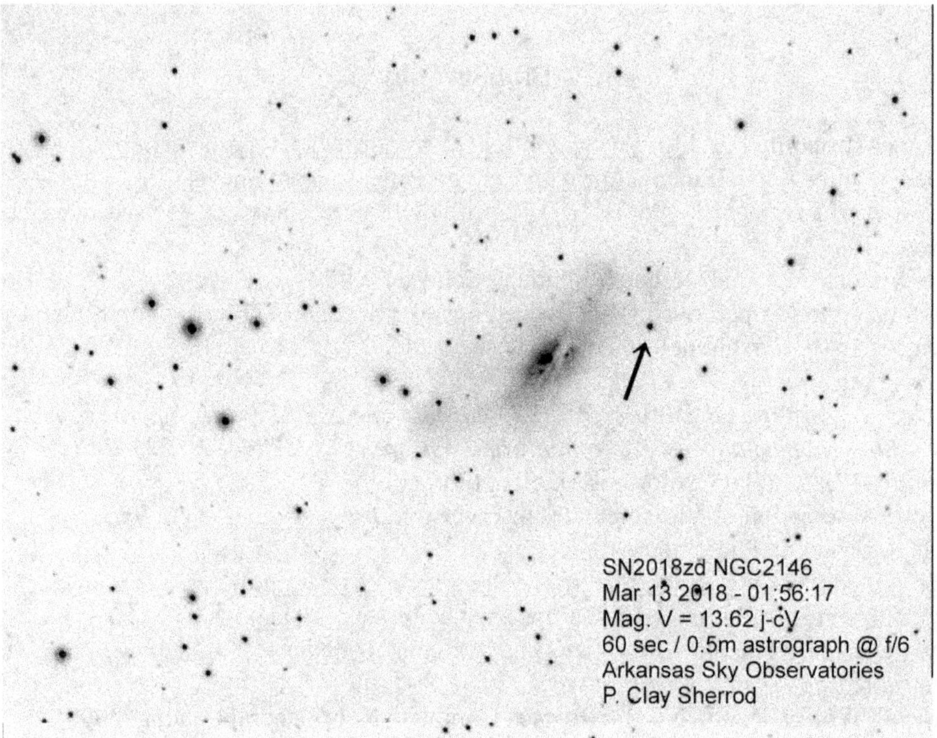

SN2018zd NGC2146
Mar 13 2018 - 01:56:17
Mag. V = 13.62 j-cV
60 sec / 0.5m astrograph @ f/6
Arkansas Sky Observatories
P. Clay Sherrod

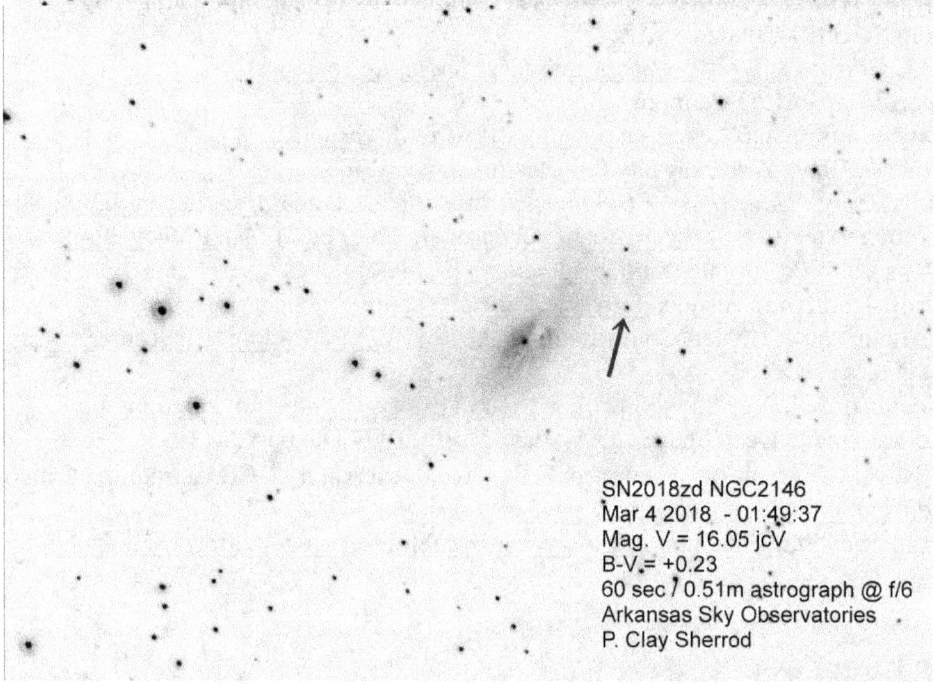

SN2018zd NGC2146
Mar 4 2018 - 01:49:37
Mag. V = 16.05 jcV
B-V = +0.23
60 sec / 0.51m astrograph @ f/6
Arkansas Sky Observatories
P. Clay Sherrod

PART II

ASO
Photographic Atlas
of
300 NGC GALAXIES

Suitable for Advanced Small Observatories
Galaxies to 15th Magnitude – Selected for Supernova Search

Listed in order of Right Ascension

NOTE:
A very few of the listed galaxies are not in the photographic Atlas due to their far southern declinations

(See listing for coordinates and magnitudes)

All photos taken at Arkansas Sky Observatories
Using the 0.51m Astrograph, f/5.9 – AstroPhysics GTO1600
Exposures = 60-seconds – Field = 15' x 20'
North UP, East to LEFT
LIMITING MAGNITUDE APPROXIMATELY **19.0**

Note that photos are presented in negative format for adequate reproduction/printing representation

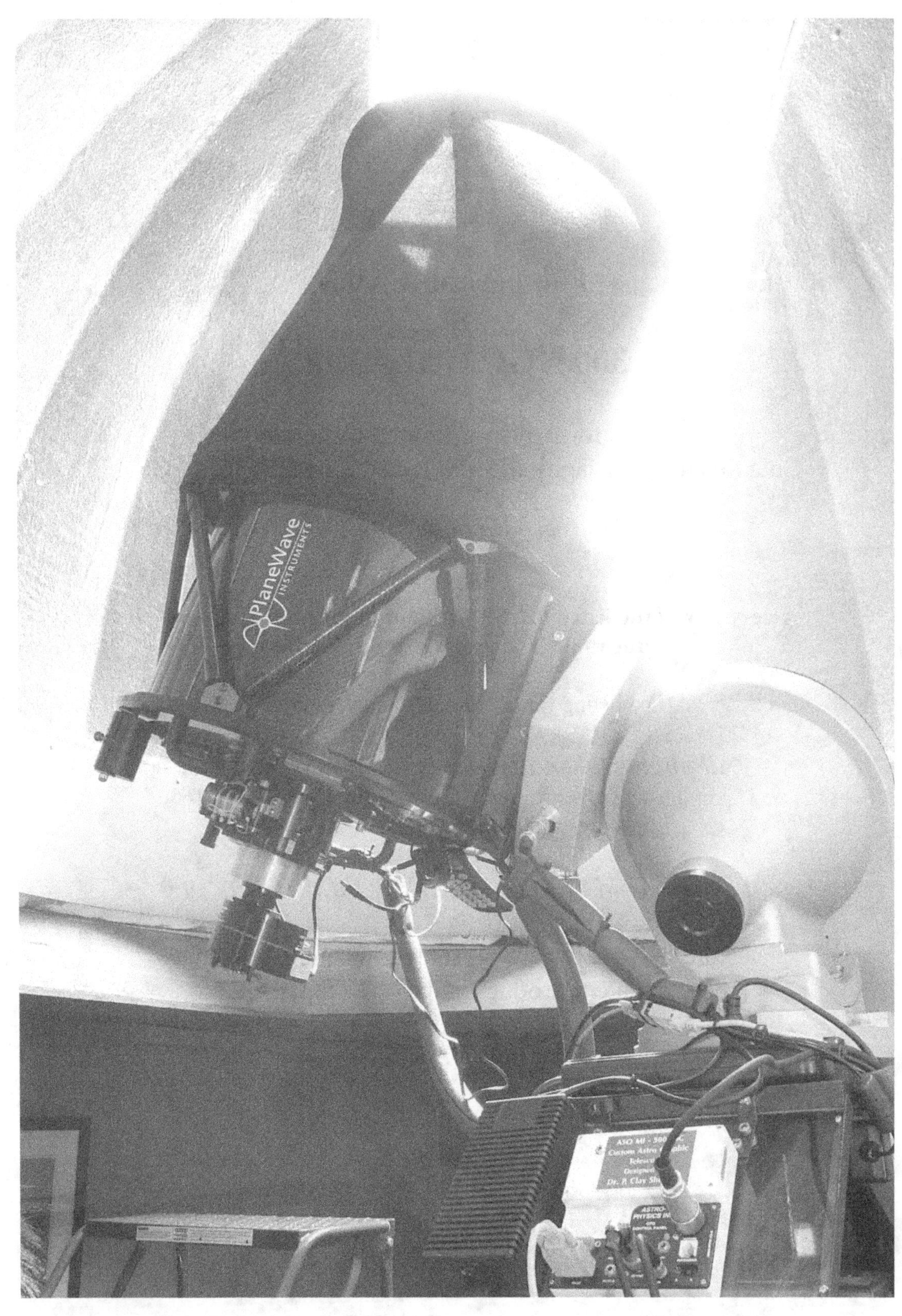
PlaneWave
INSTRUMENTS

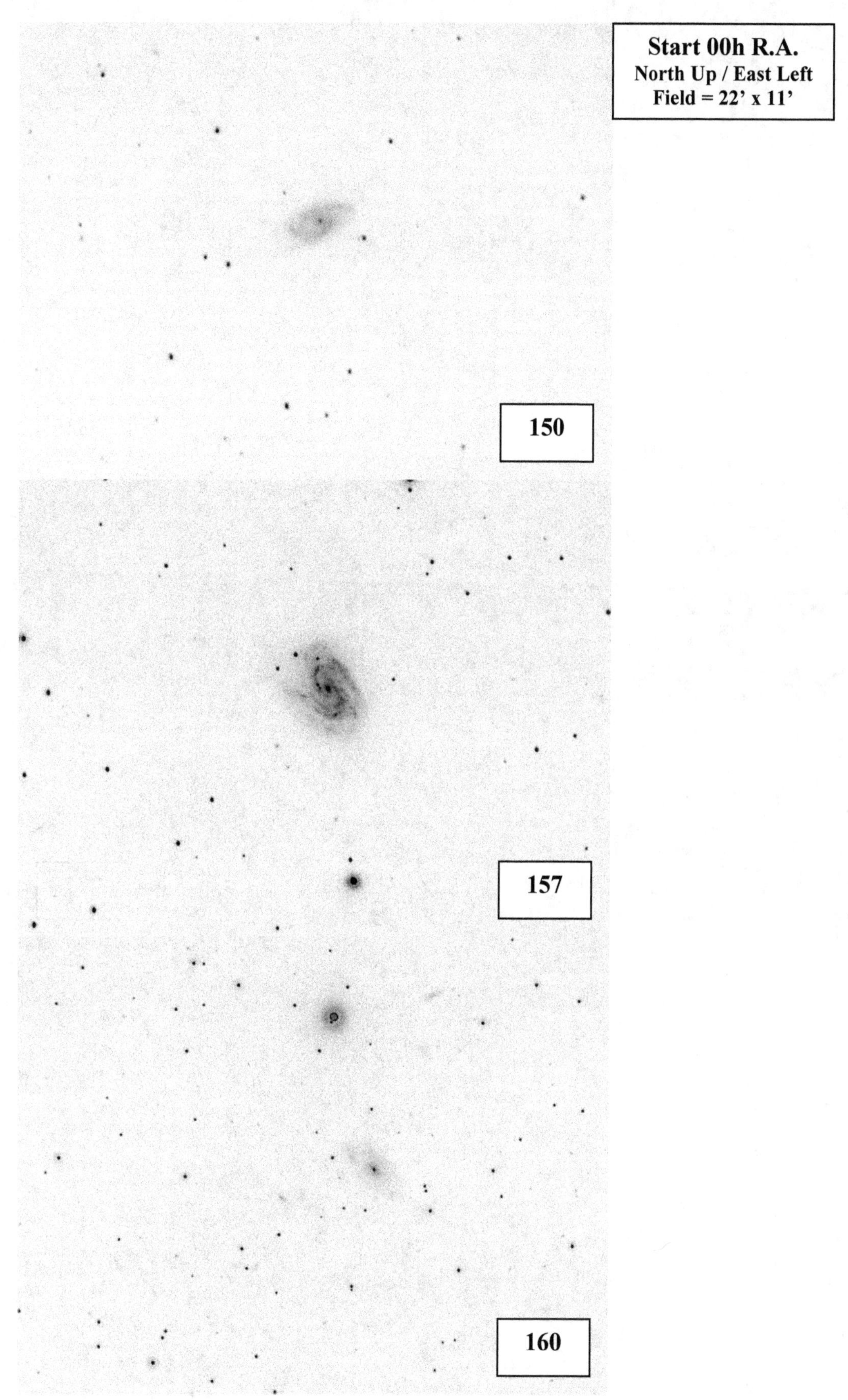
Start 00h R.A.
North Up / East Left
Field = 22' x 11'
150
157
160

RA 00h

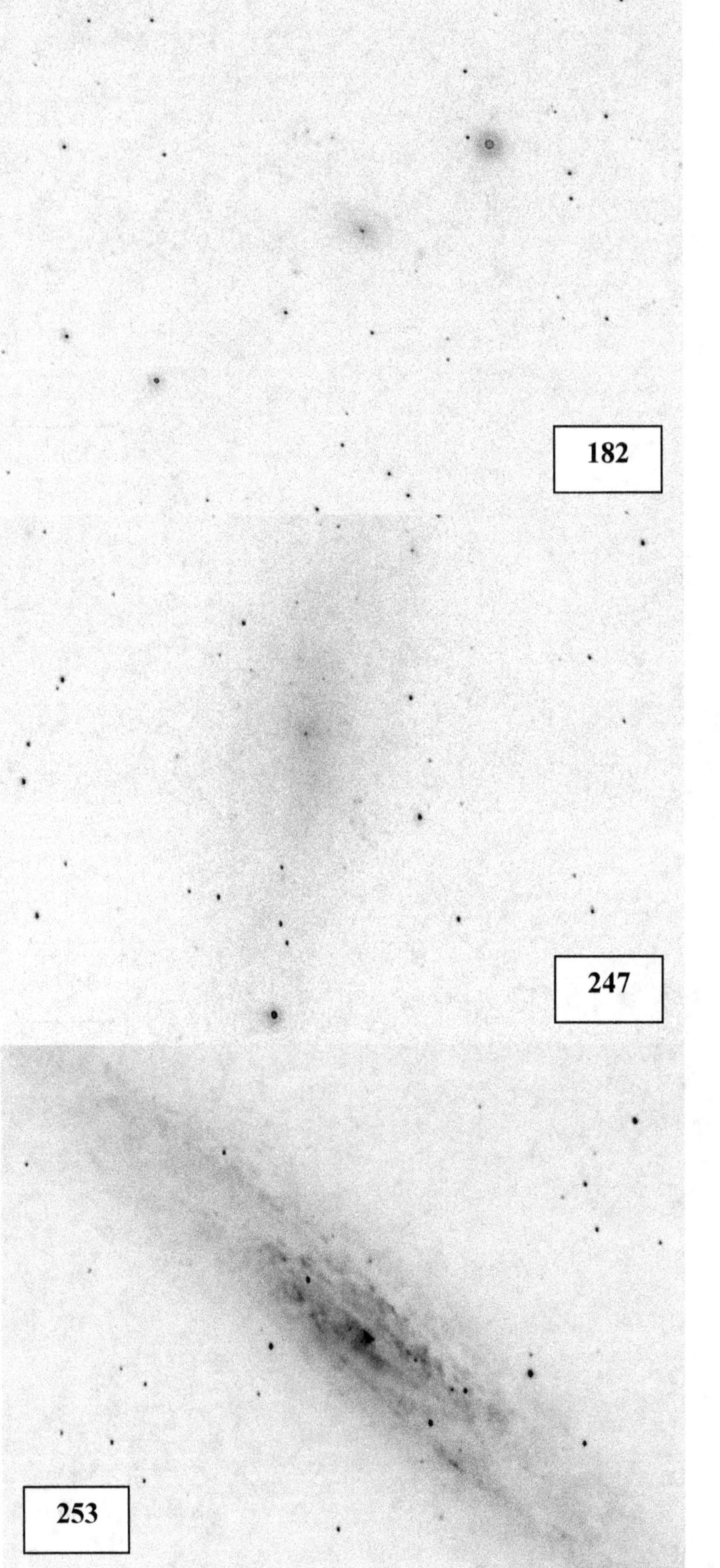

RA 00h

309

327-329-325

7814

RA 00h

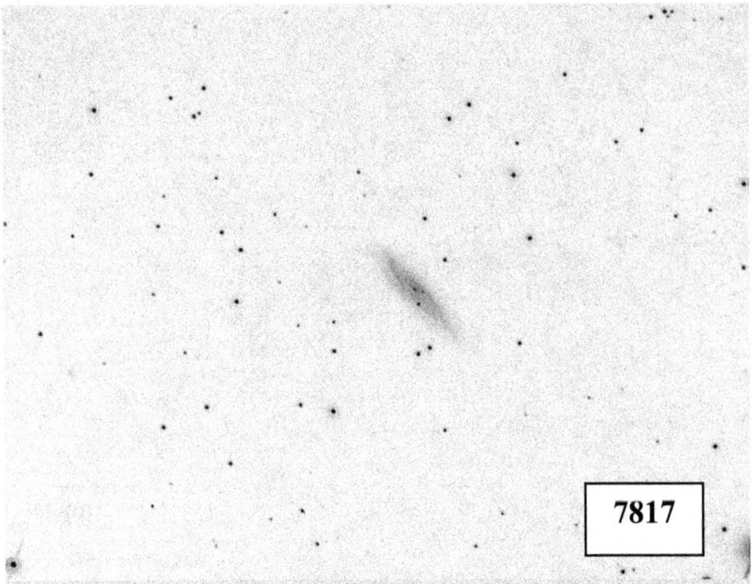

7817

Start 01h R.A.
North Up / East Left
Field = 22' x 11'

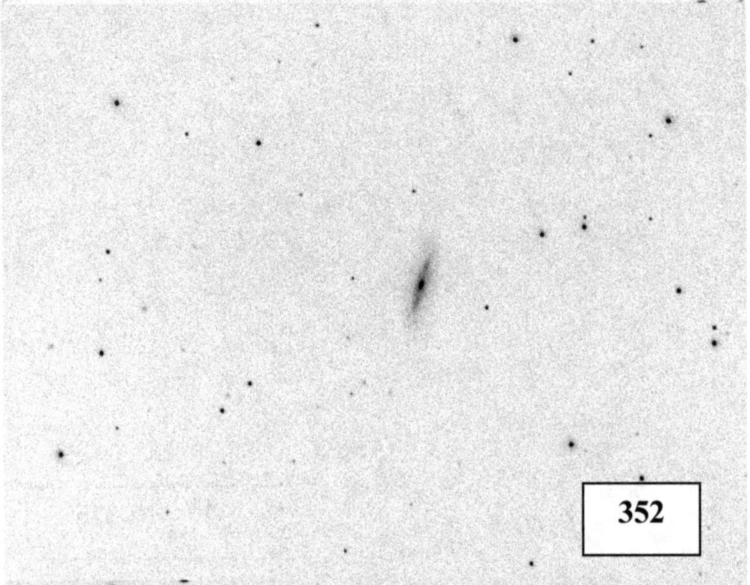

352

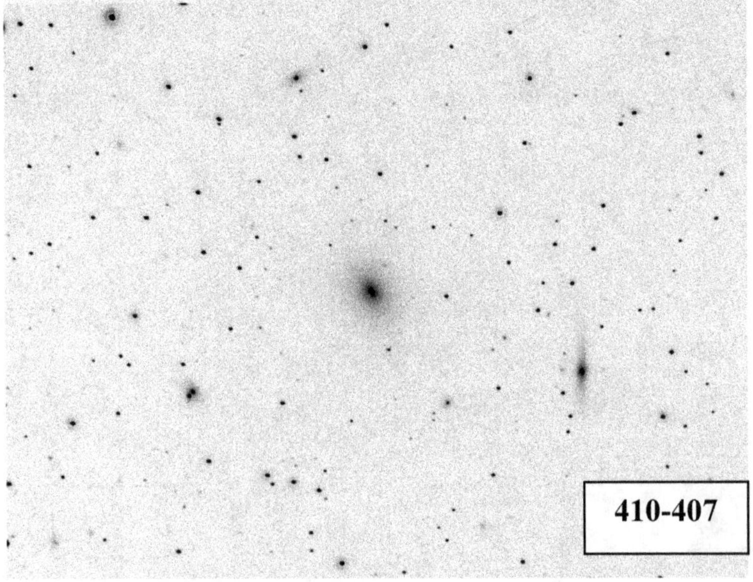

410-407

RA 01h

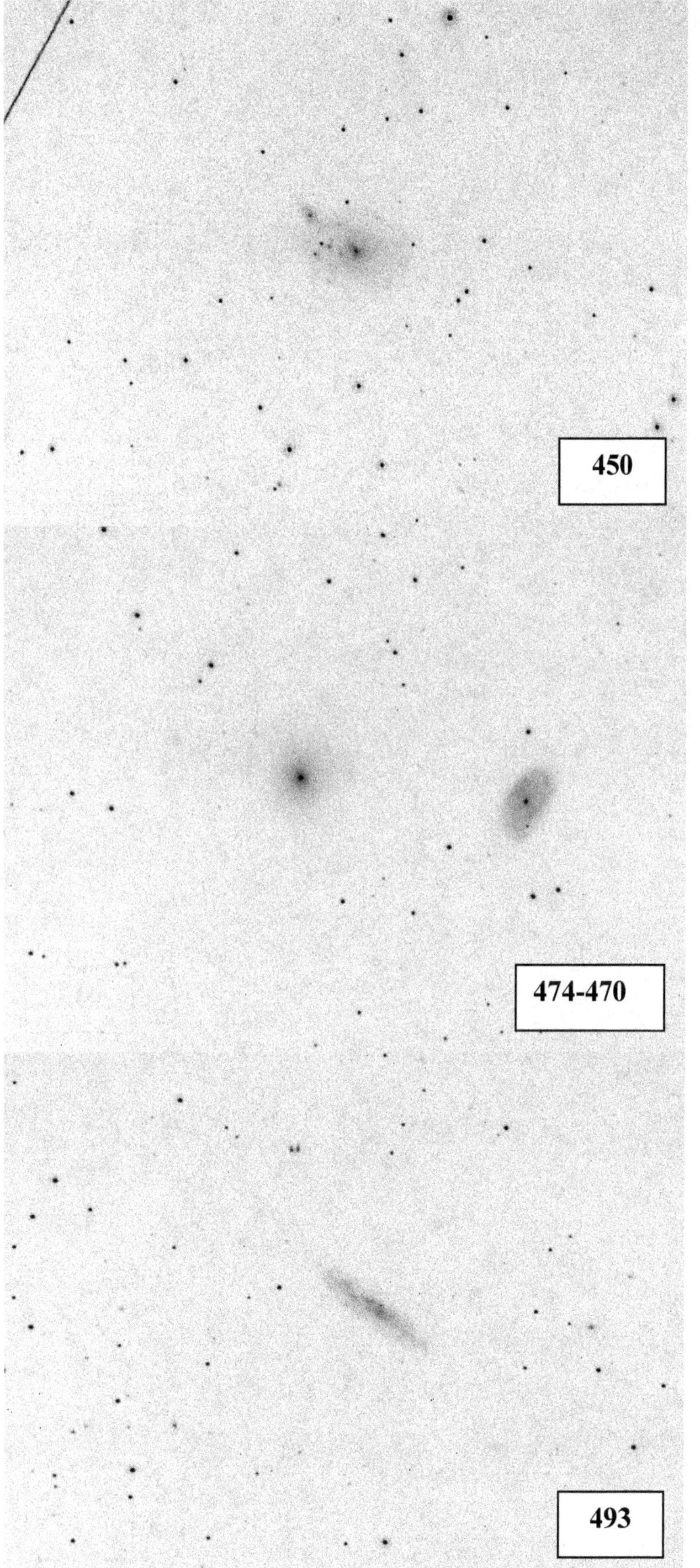

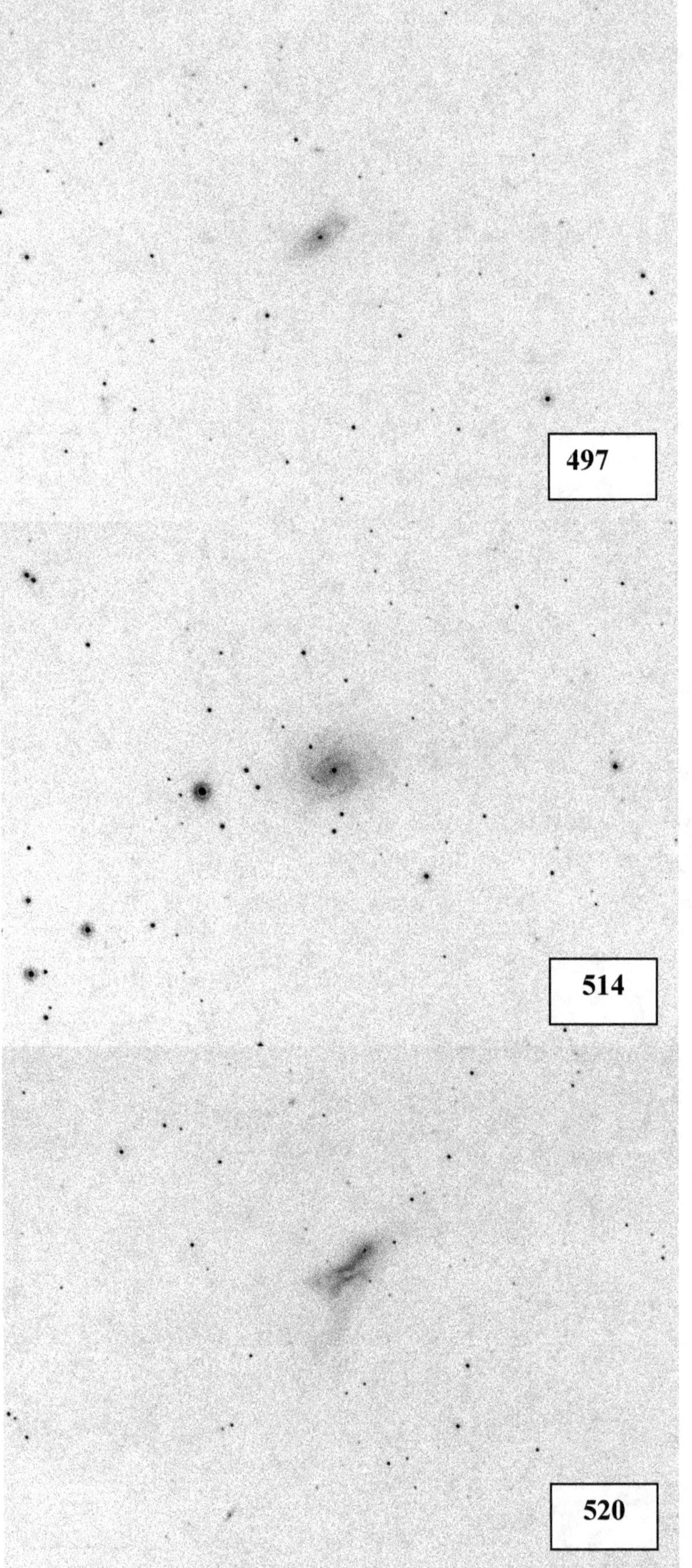
497
514
520

RA 01h

521

522

532

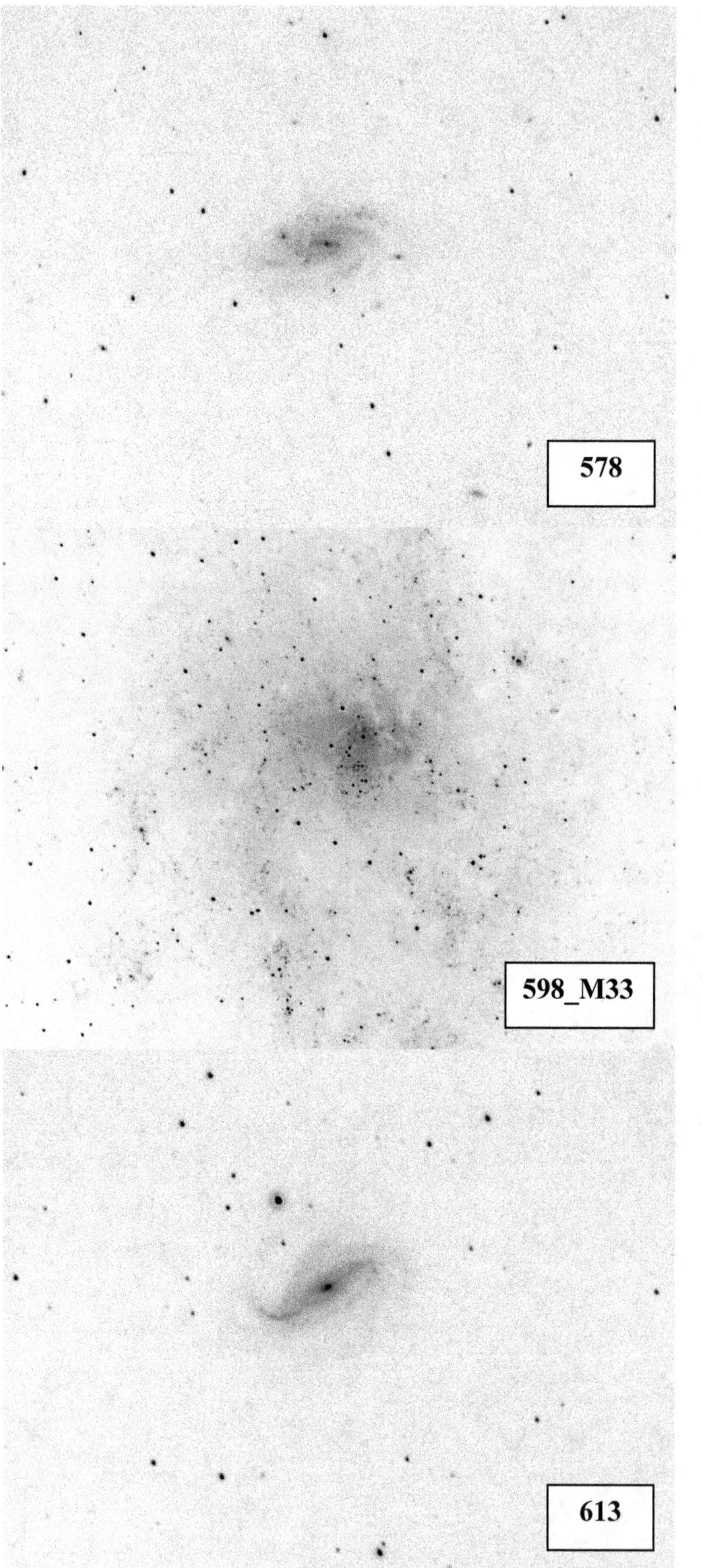
578
598_M33
613

RA 01h

615

628 – M74

660

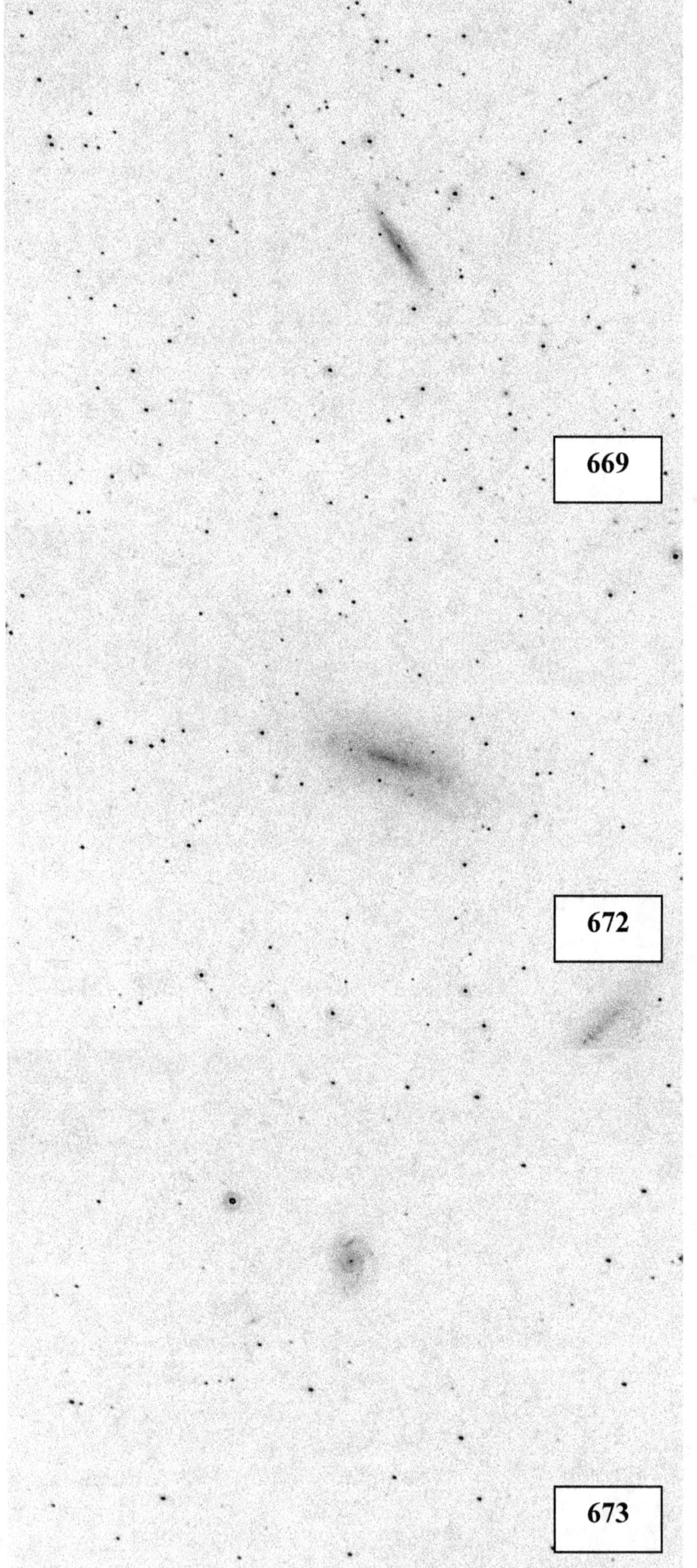

RA 01h

RA 01h

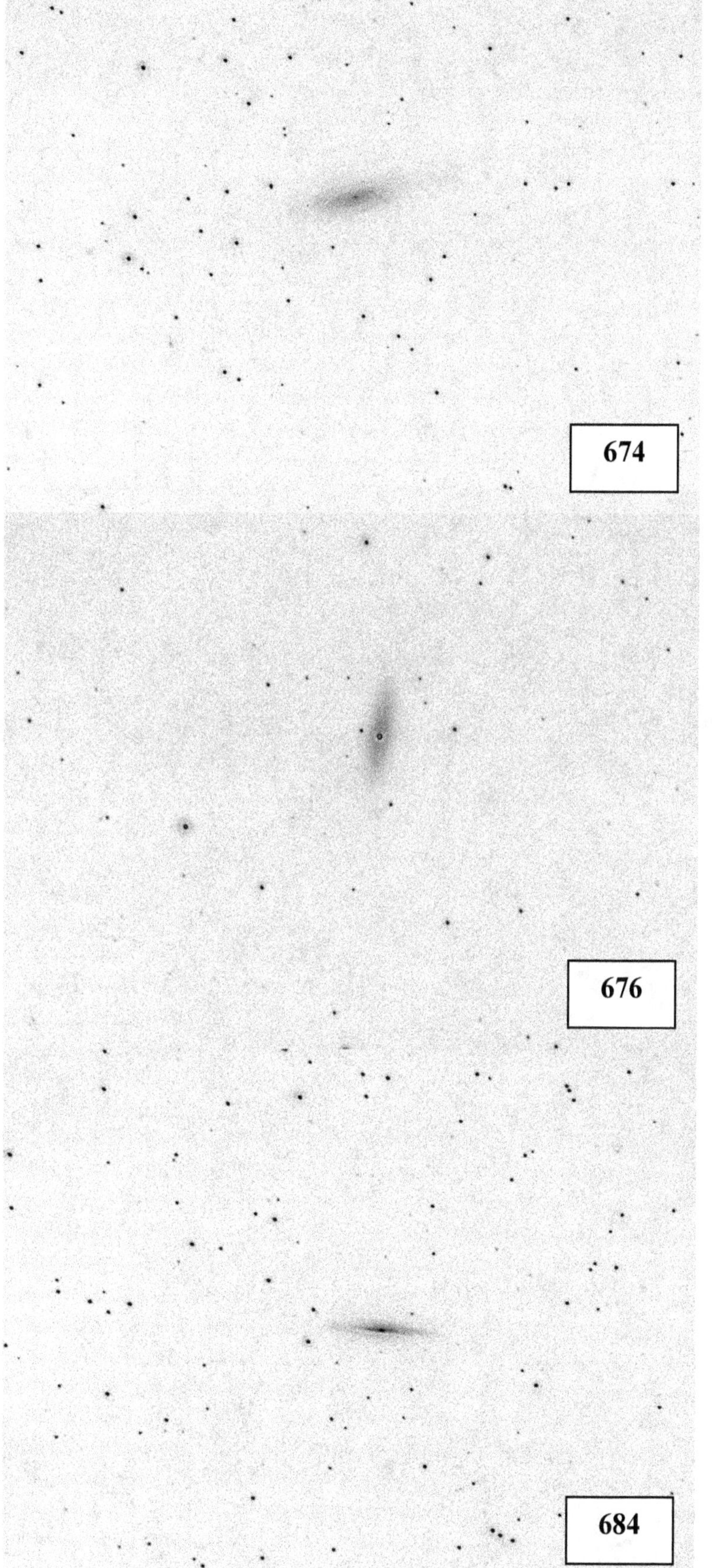

RA 01h

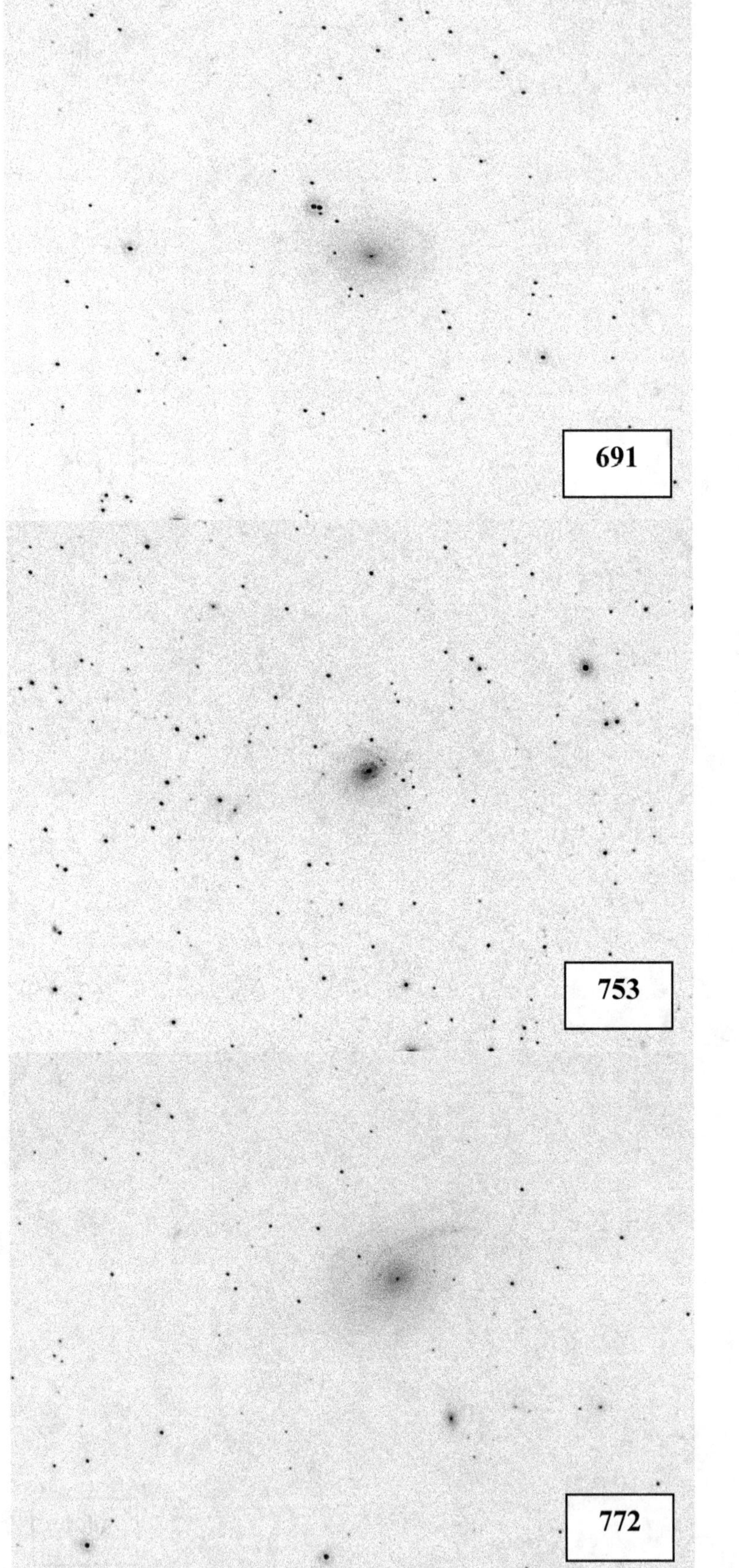

RA 01h

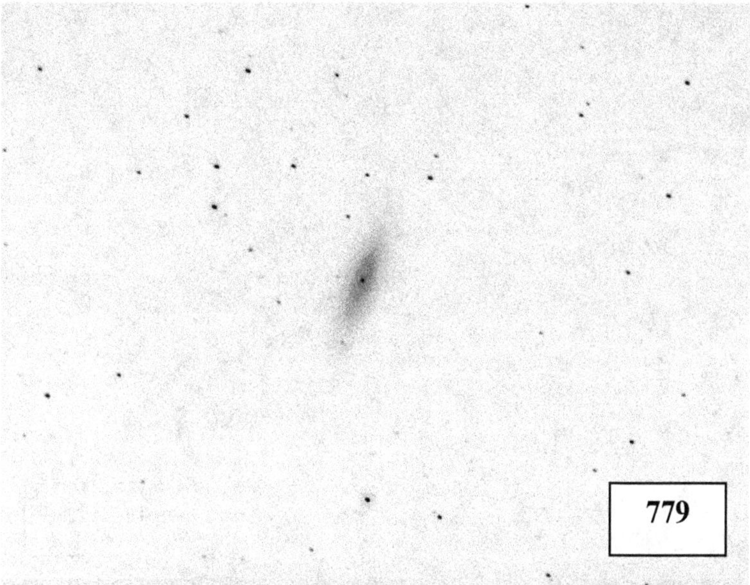

779

Start 02h R.A.
North Up / East Left
Field = 22' x 11'

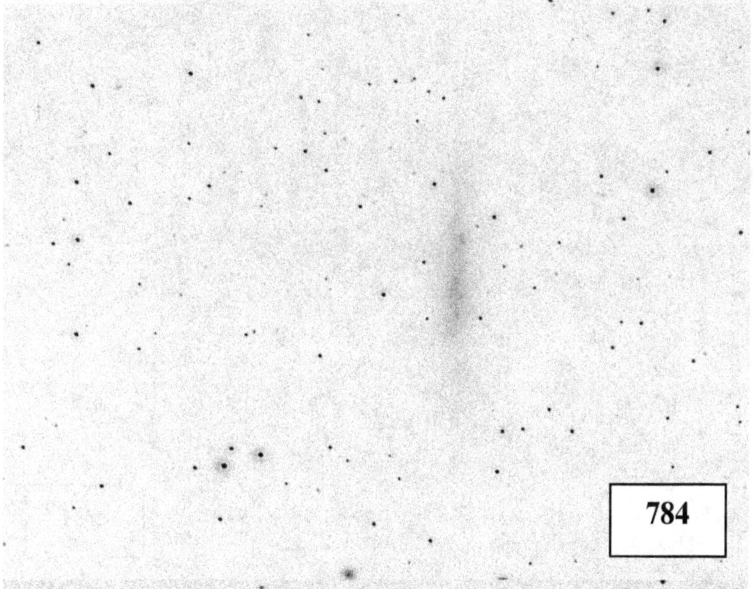

784

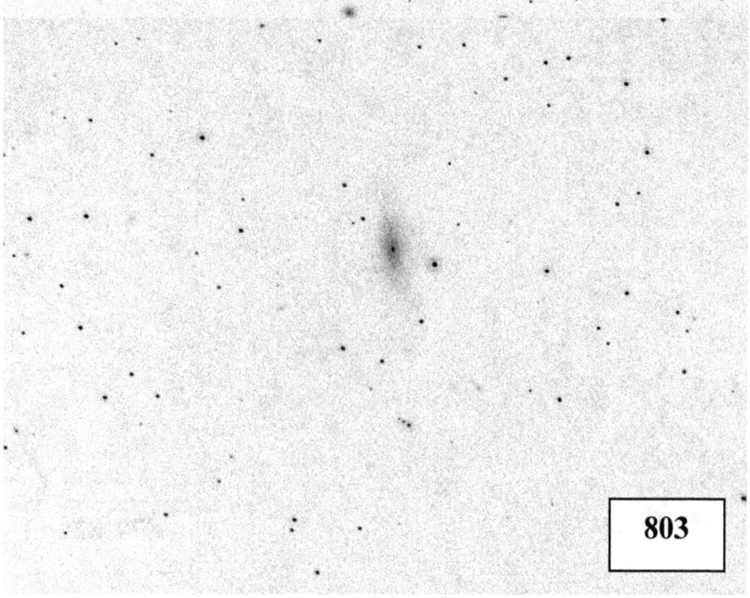

803

RA 02h

818

864

877-876

RA 02h

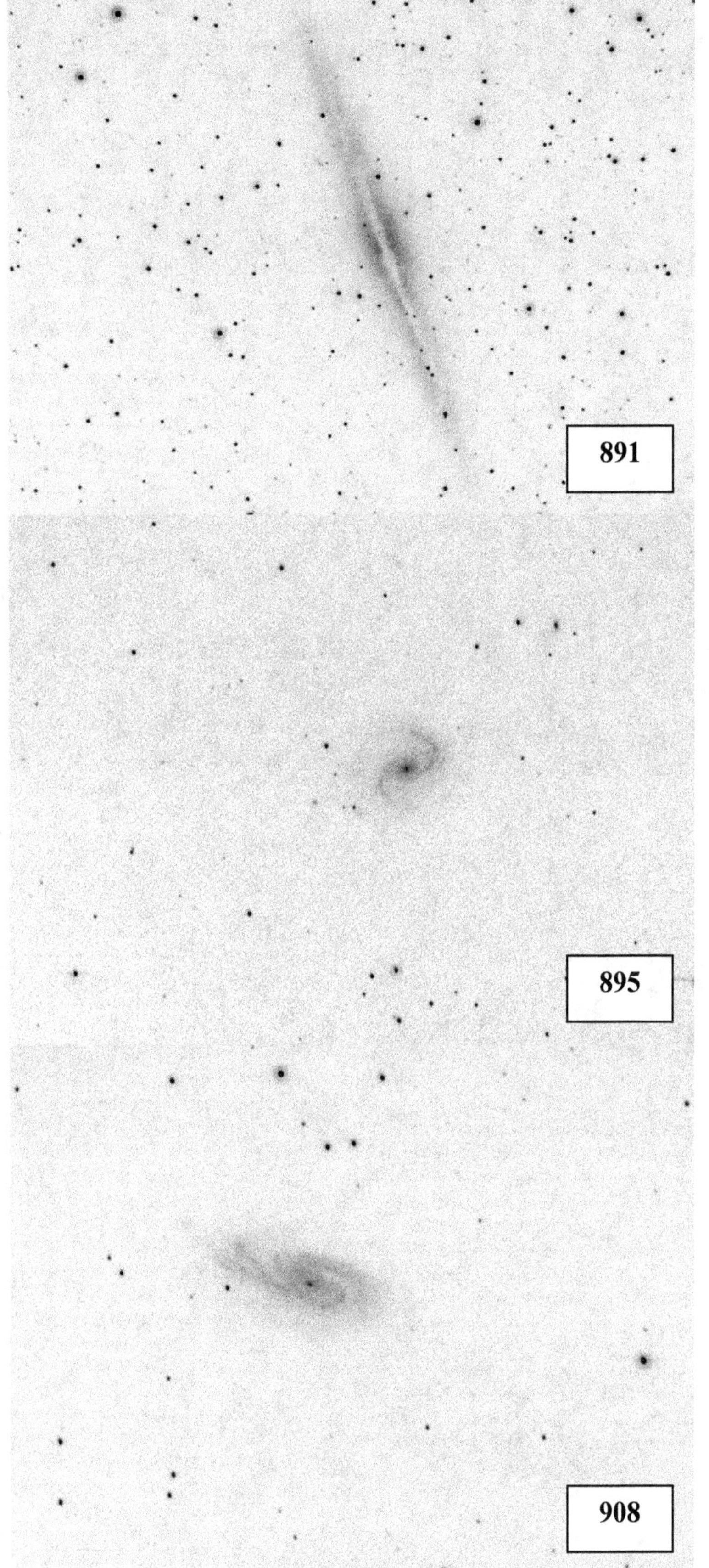

RA 02h

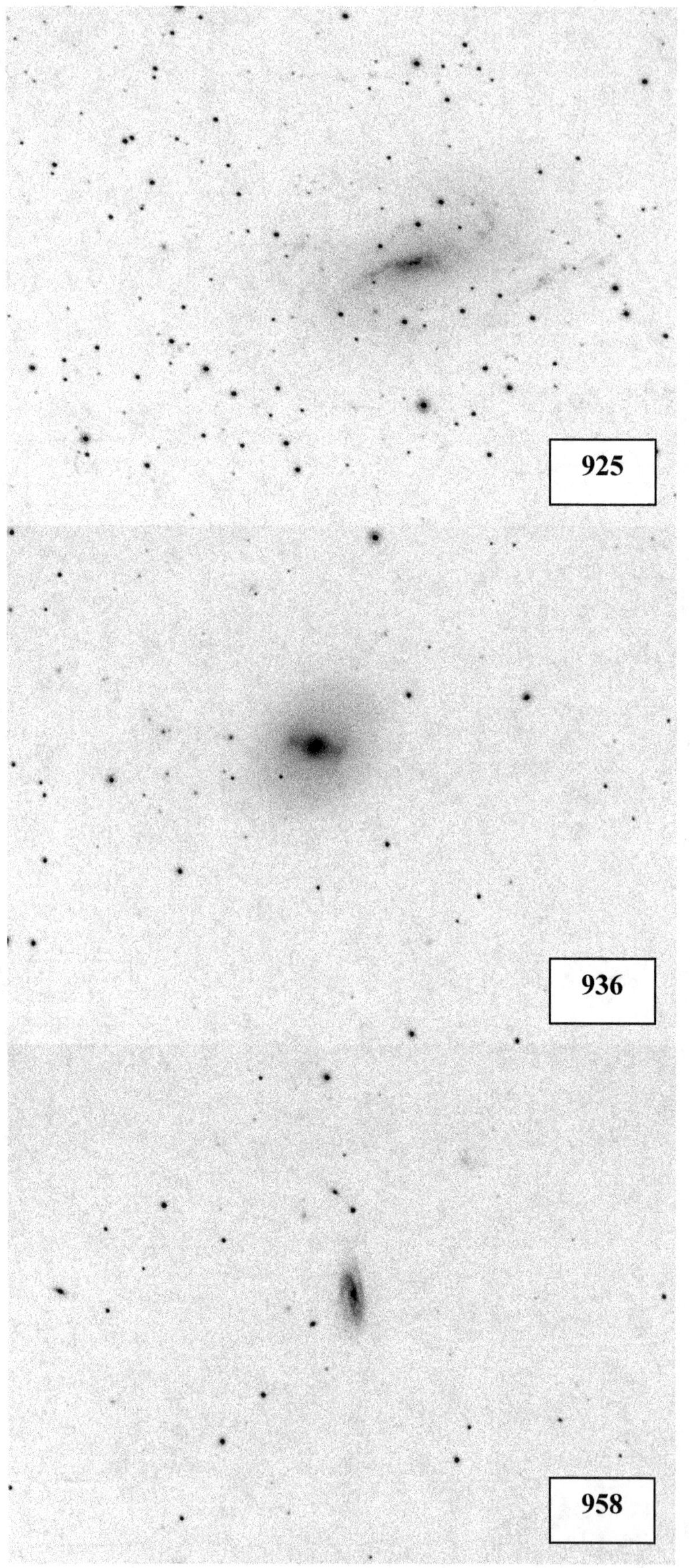

RA 02h

972

1003

1023

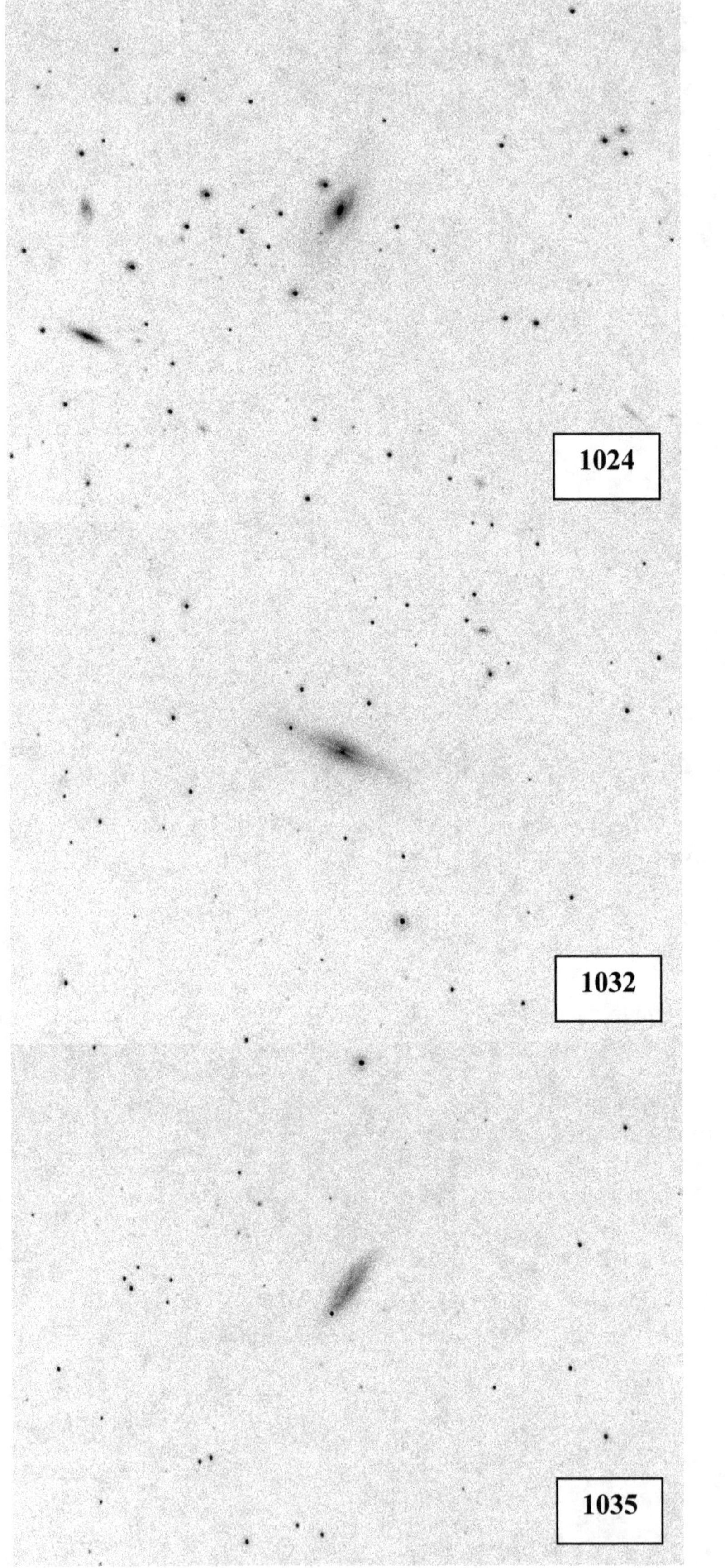
1024
1032
1035

RA 02h

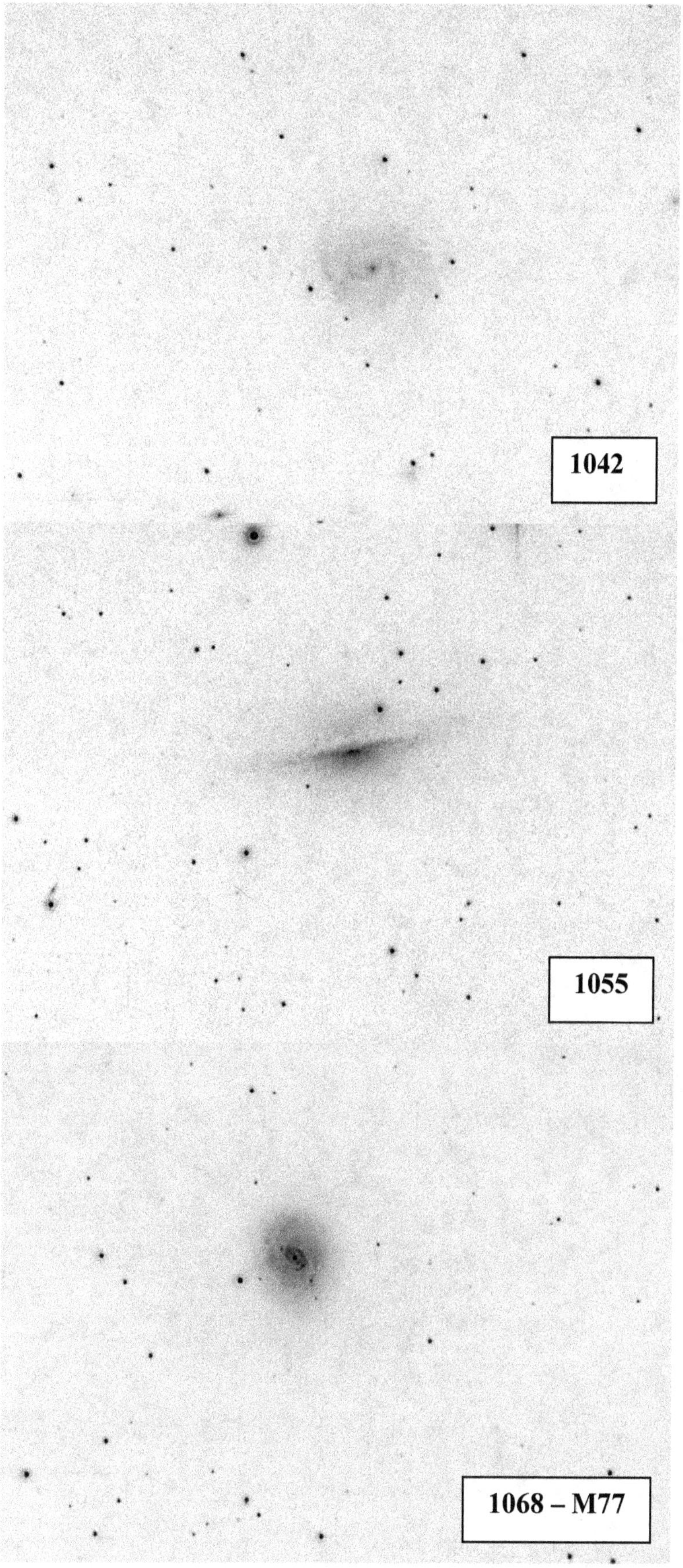

1042

1055

1068 – M77

RA 02h

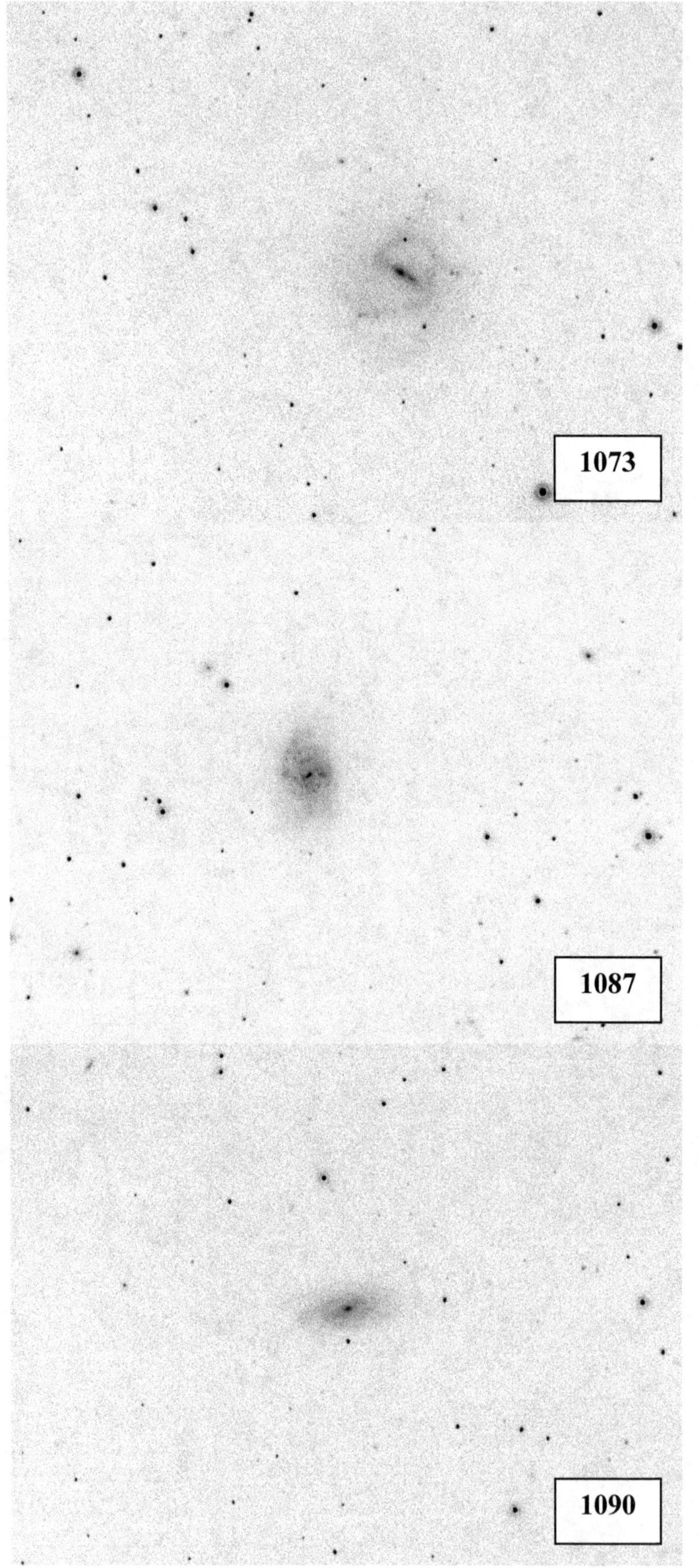

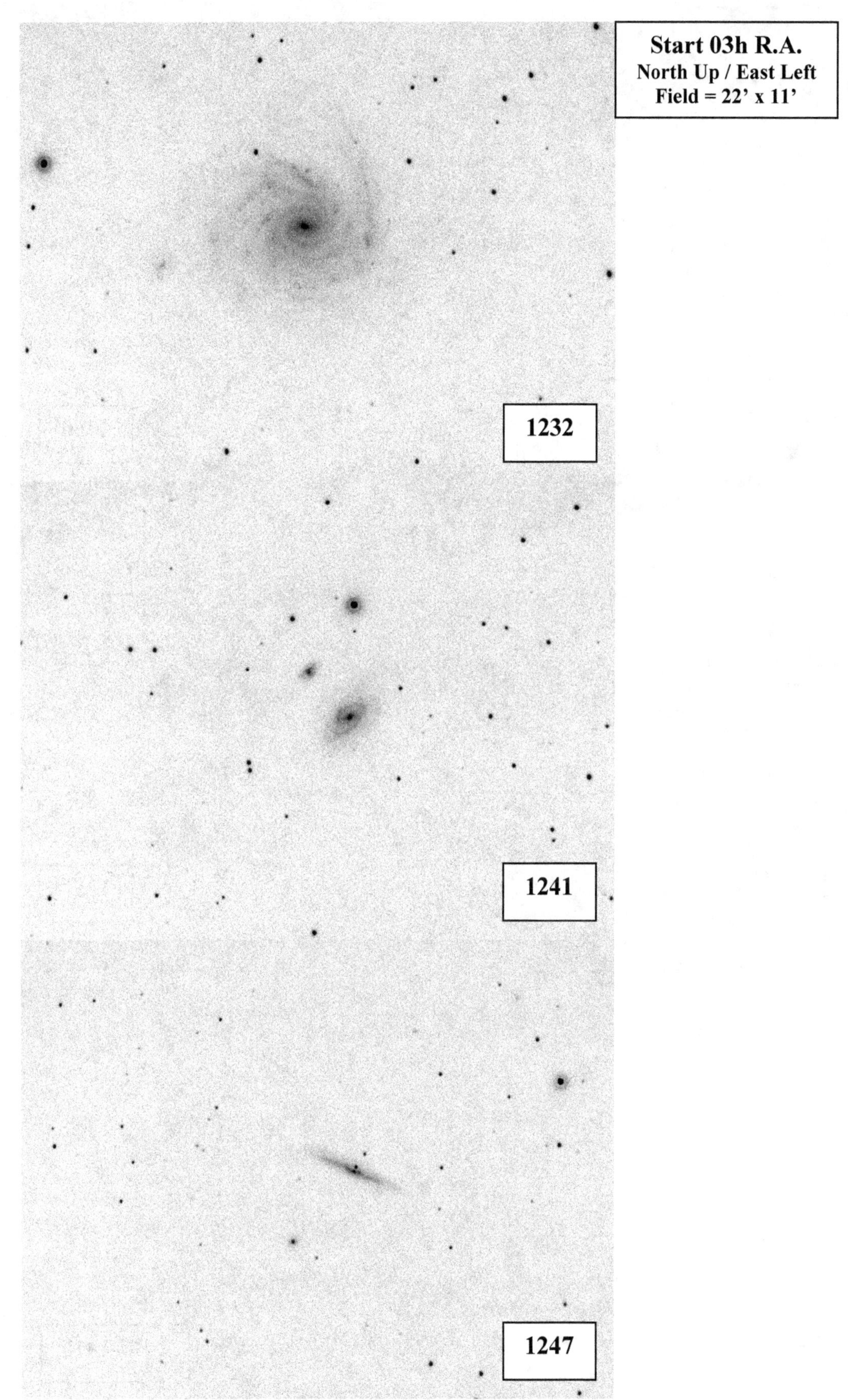
Start 03h R.A.
North Up / East Left
Field = 22' x 11'
1232
1241
1247

RA 03h

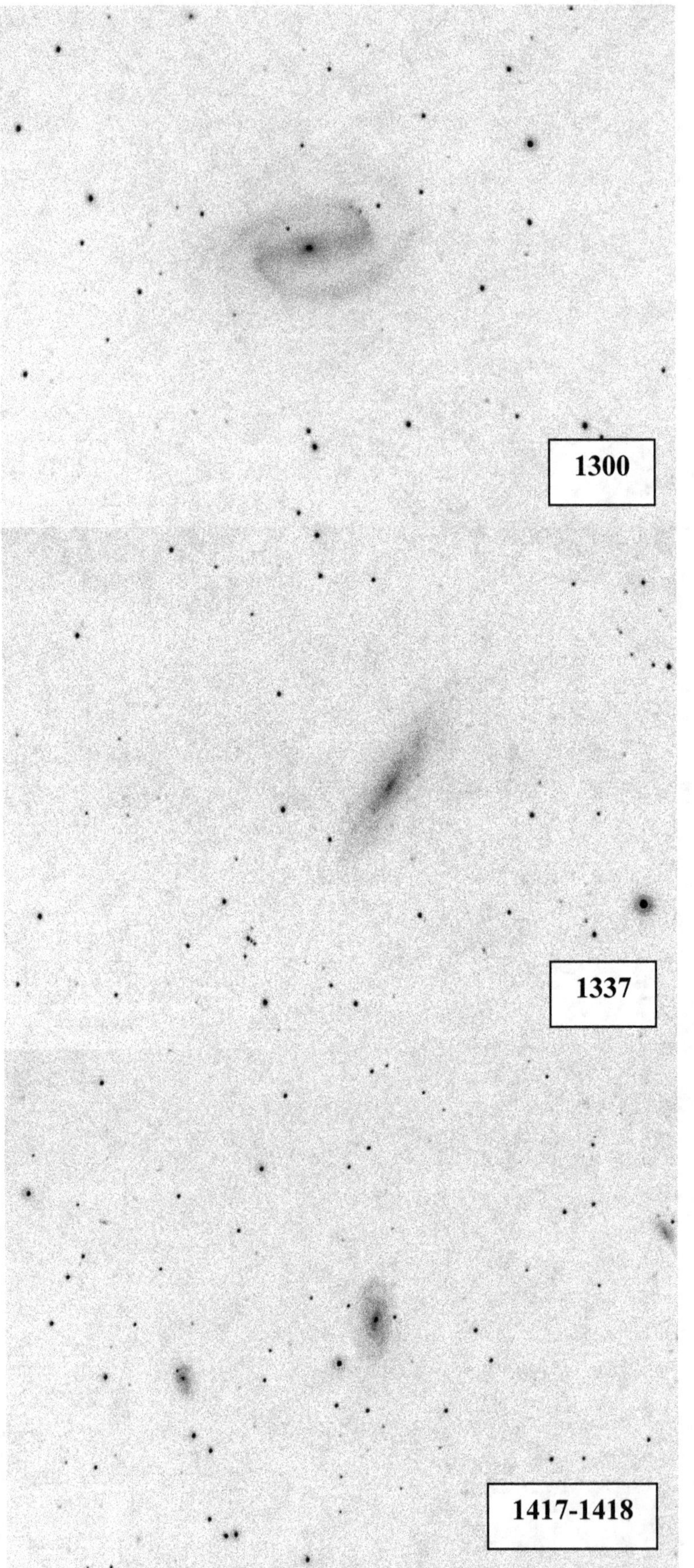

RA 03h

1421

Start 04h R.A.
North Up / East Left
Field = 22' x 11'

1507

1530

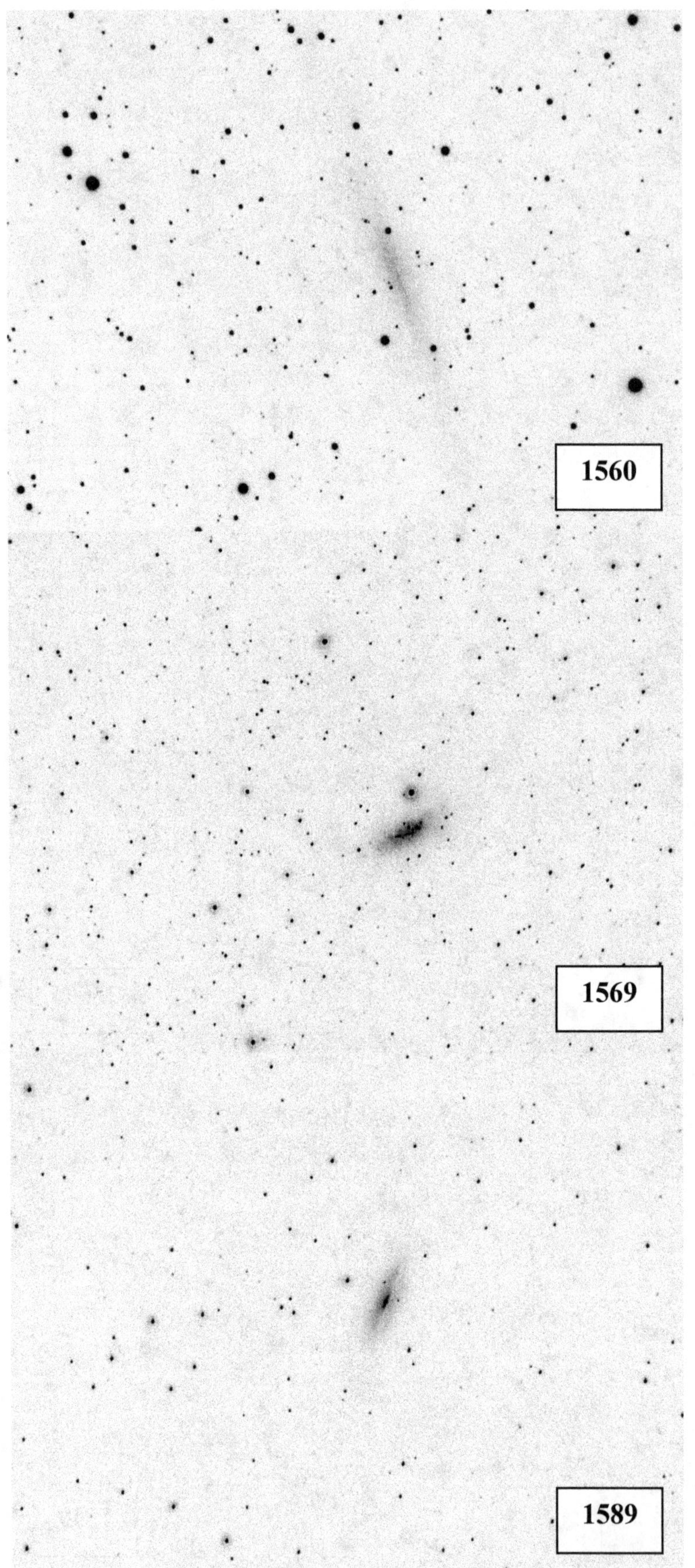

RA 04h

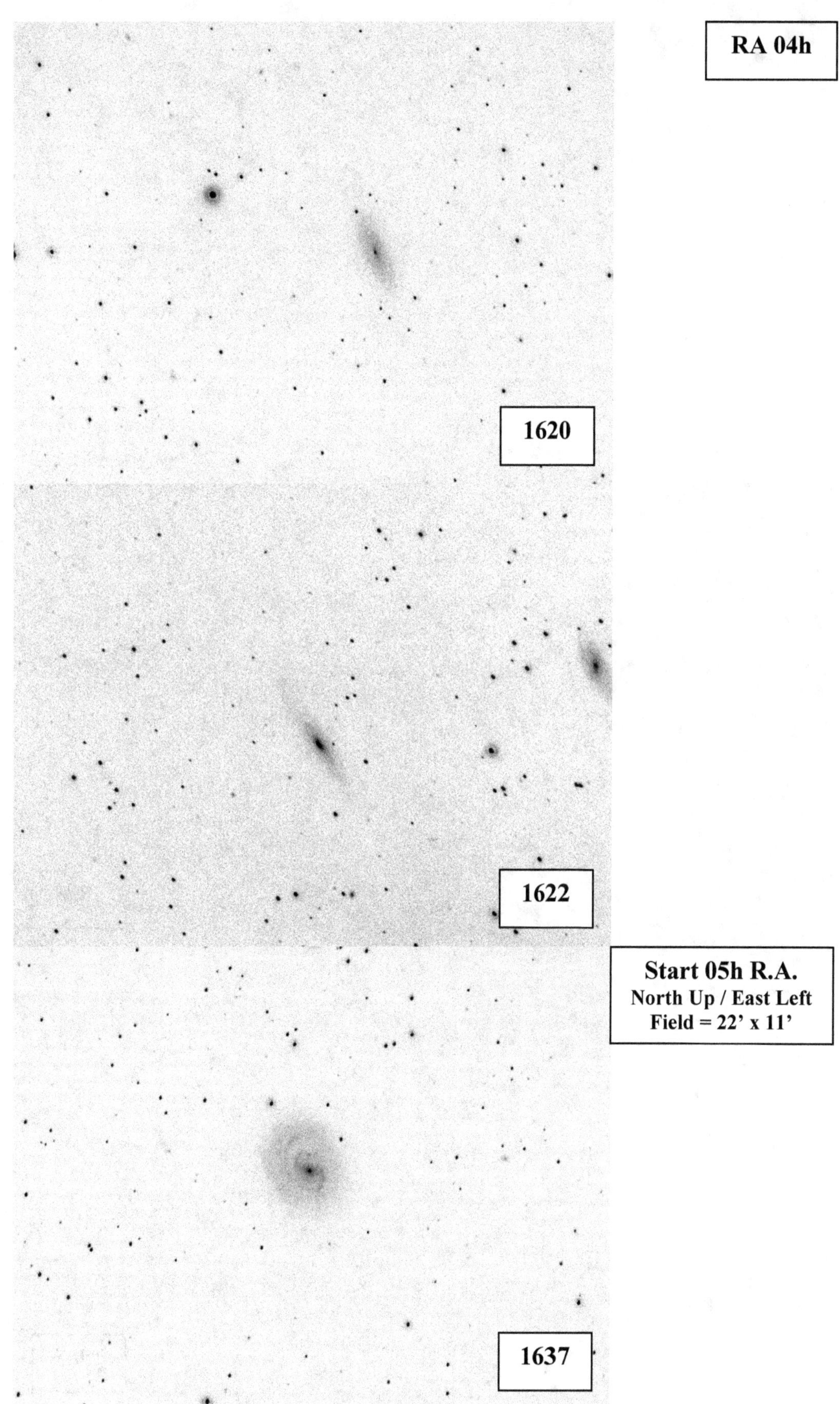
RA 04h
1620
1622
Start 05h R.A.
North Up / East Left
Field = 22' x 11'
1637

RA 05h

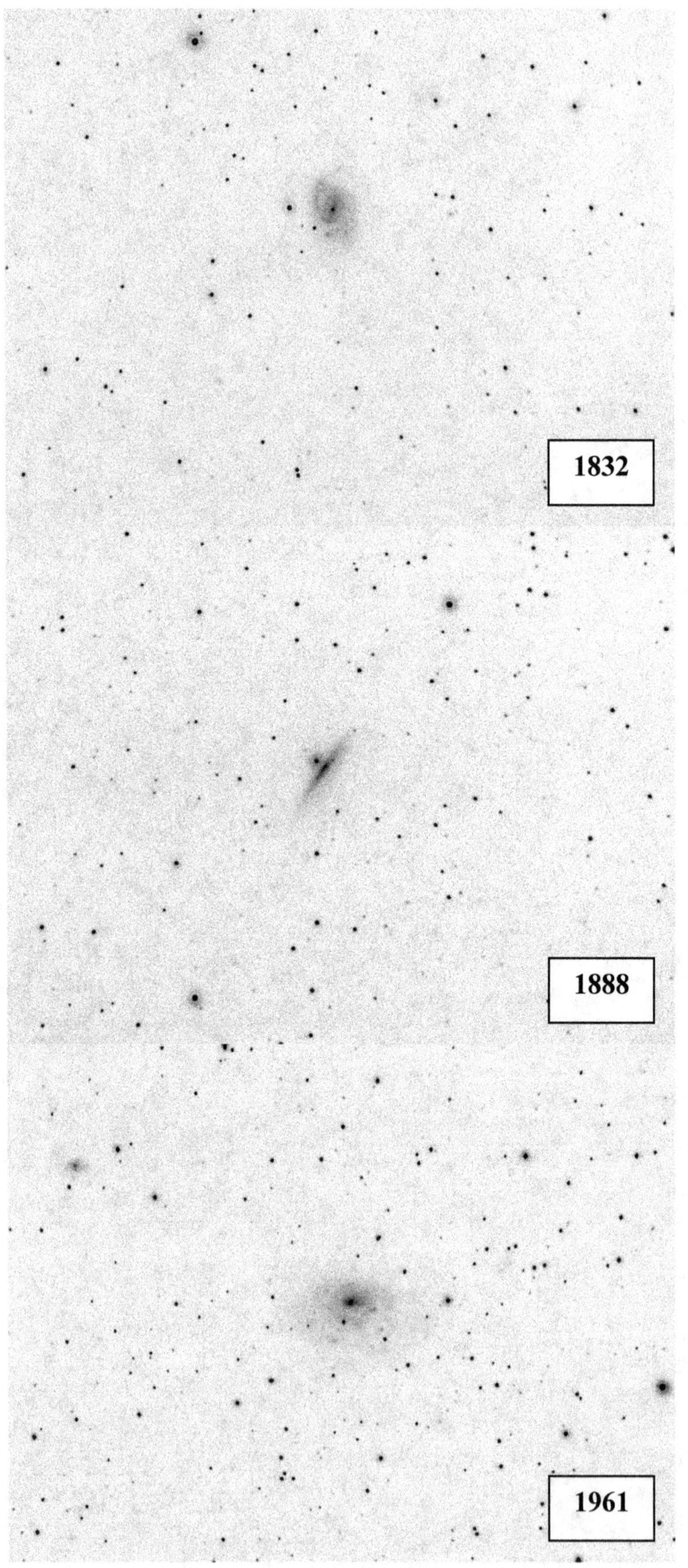

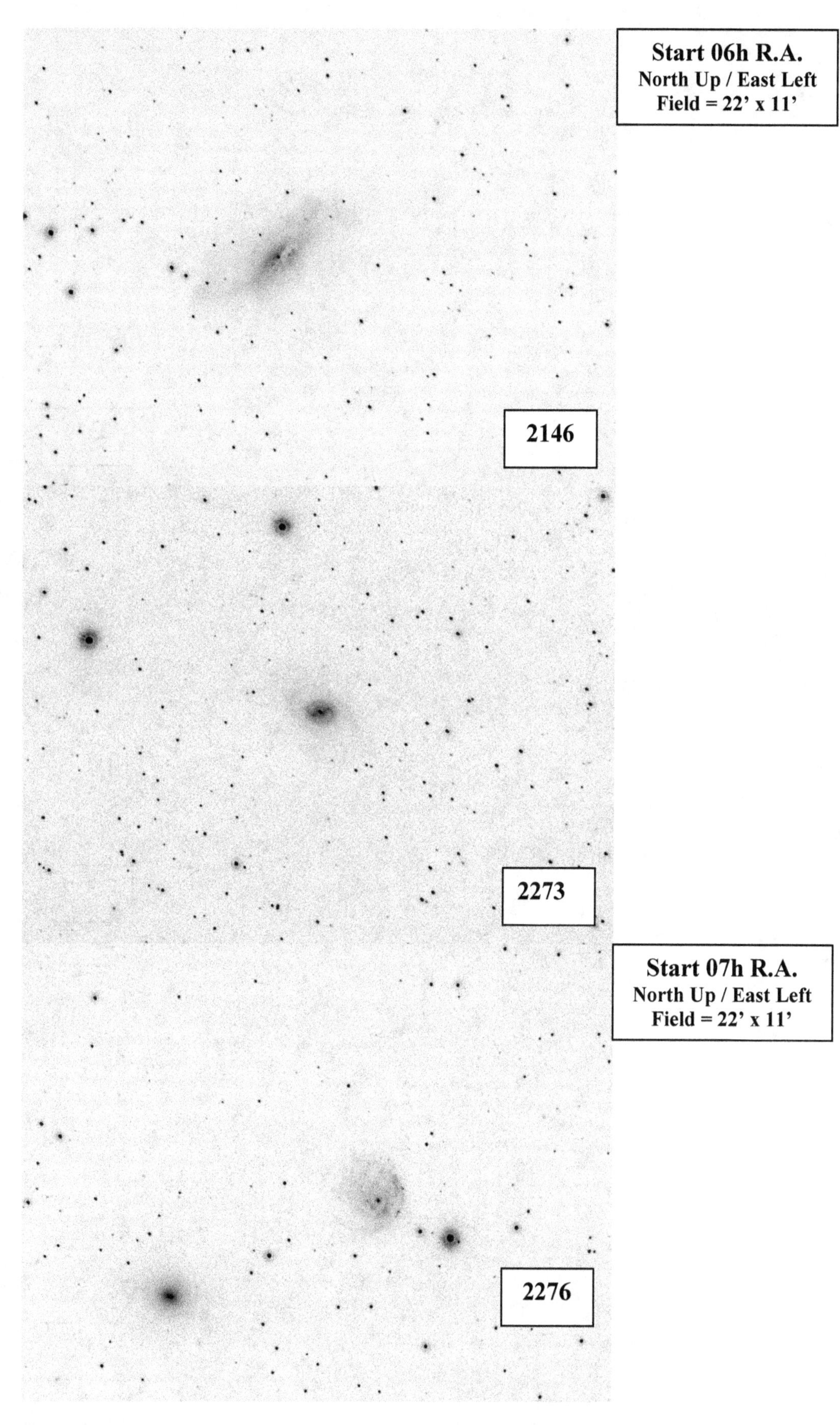
Start 06h R.A.
North Up / East Left
Field = 22' x 11'
2146
2273
Start 07h R.A.
North Up / East Left
Field = 22' x 11'
2276

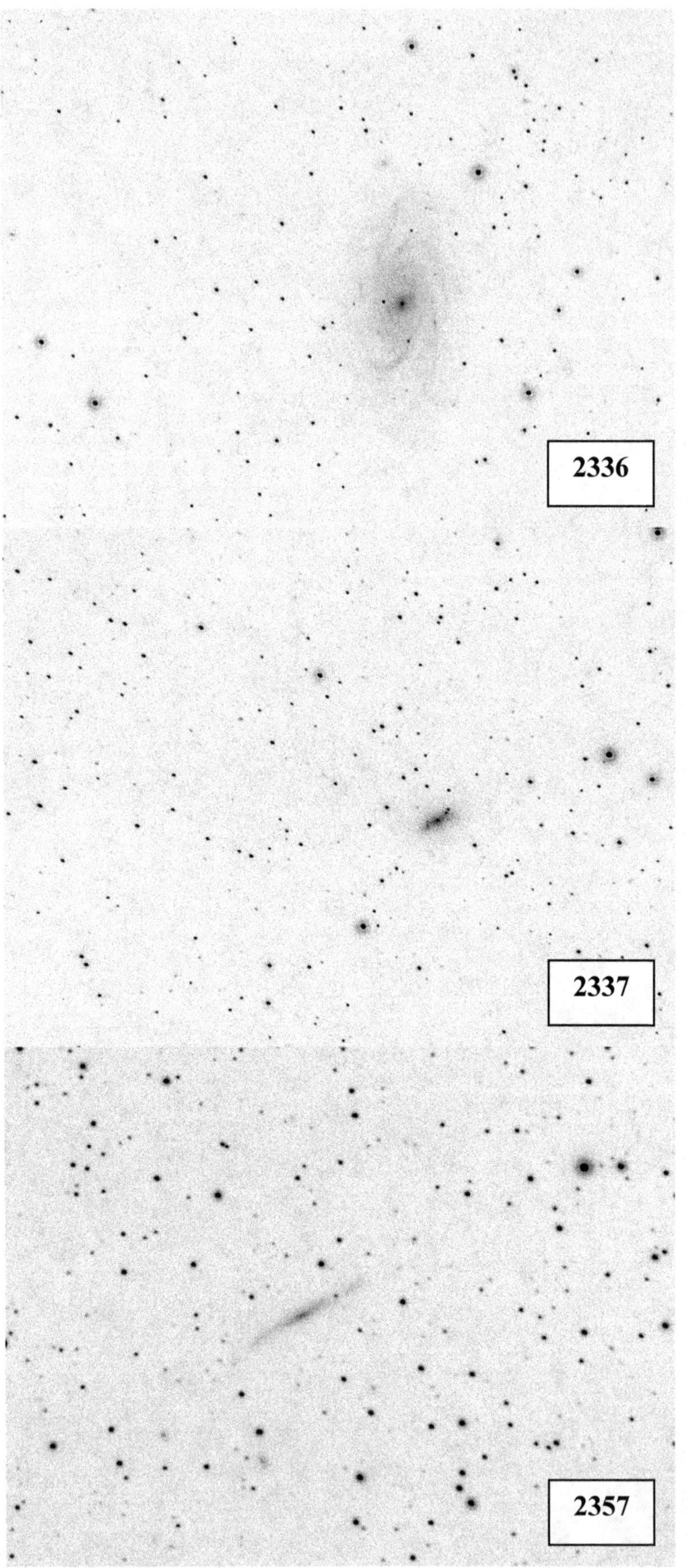
2336
2337
2357

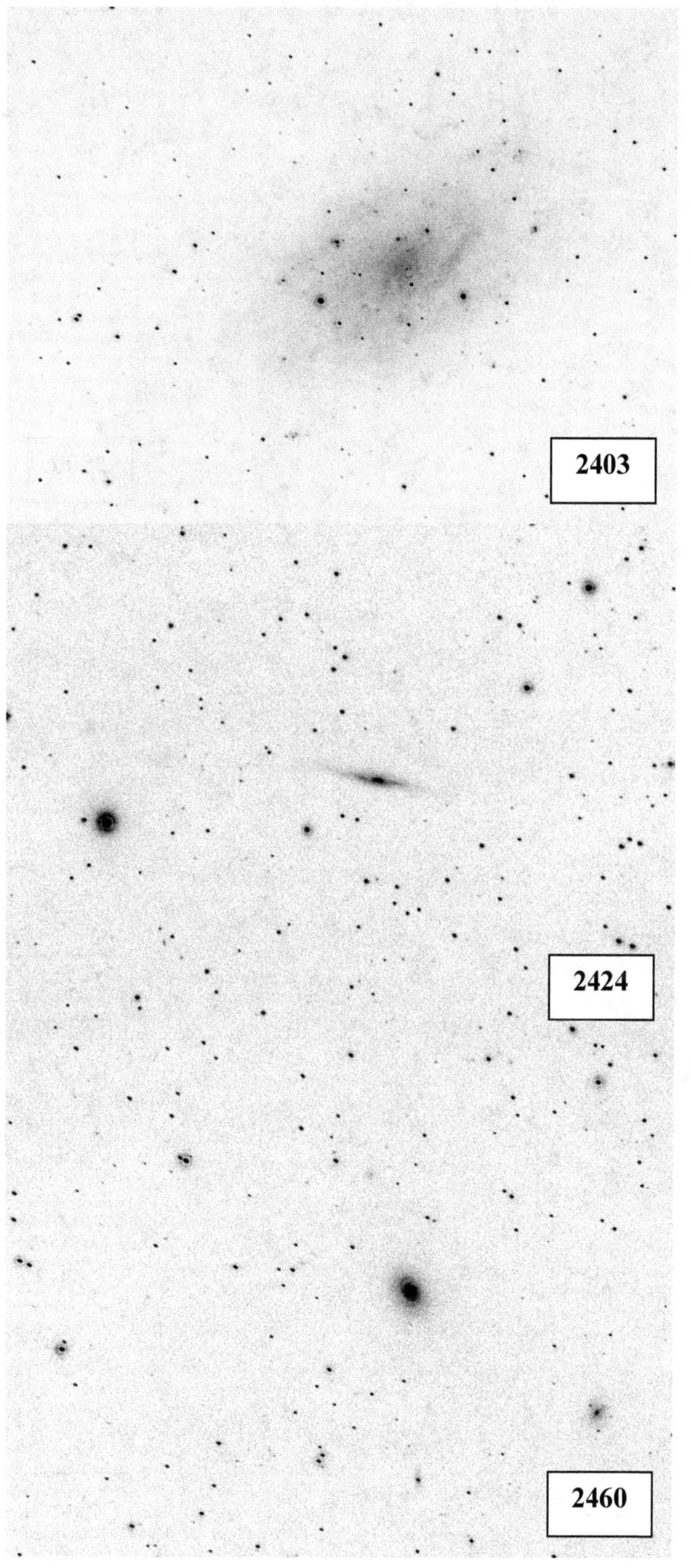
2403
2424
2460

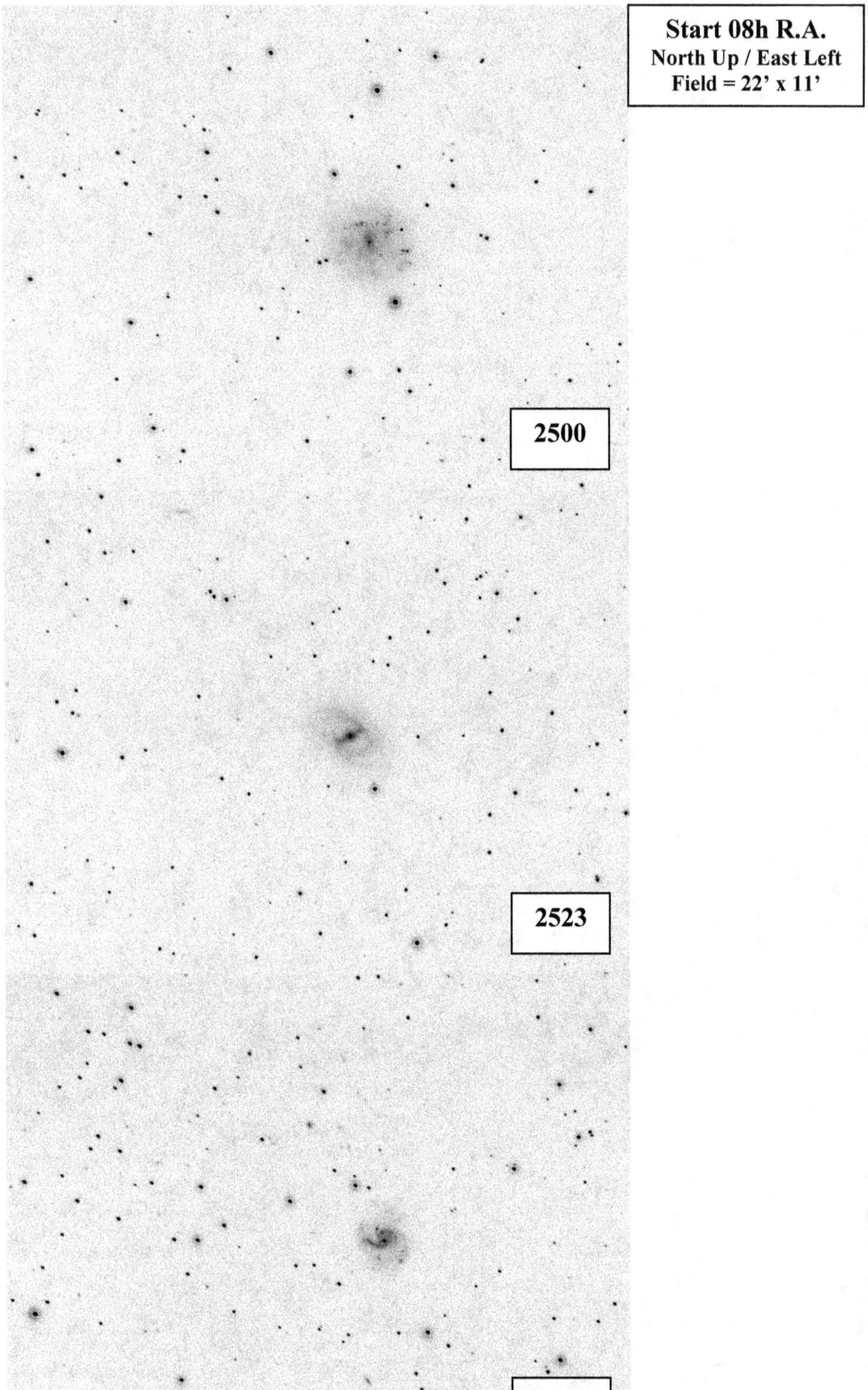

Start 08h R.A.
North Up / East Left
Field = 22' x 11'

2500

2523

2532

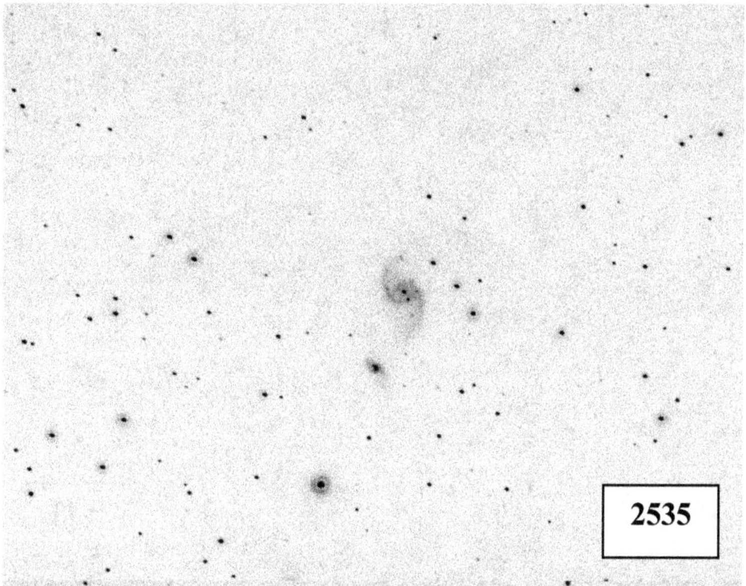

RA 08h

2535

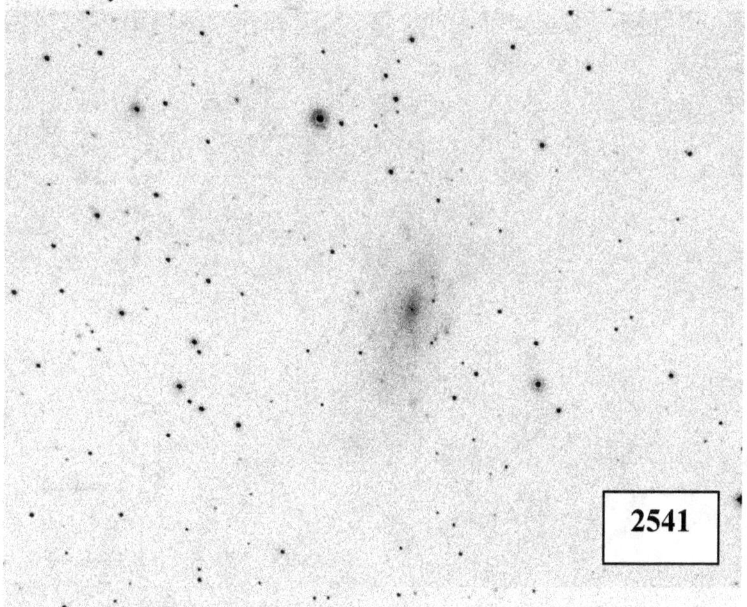

2541

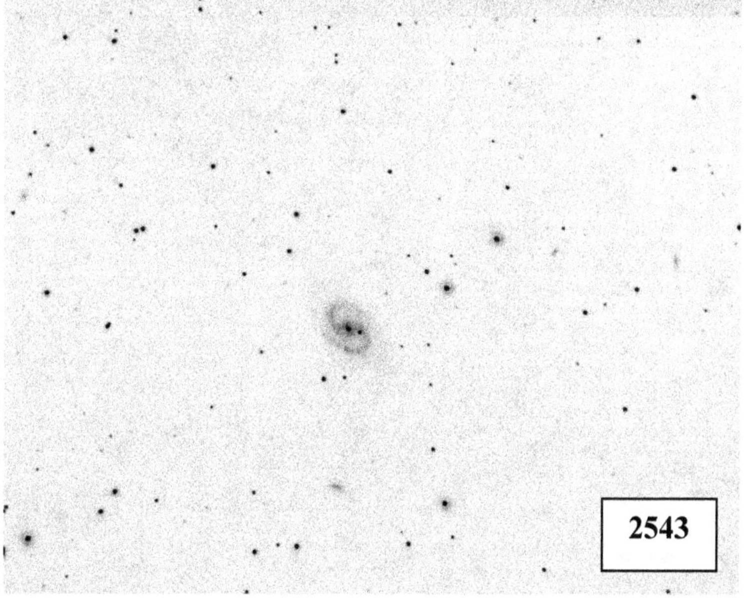

2543

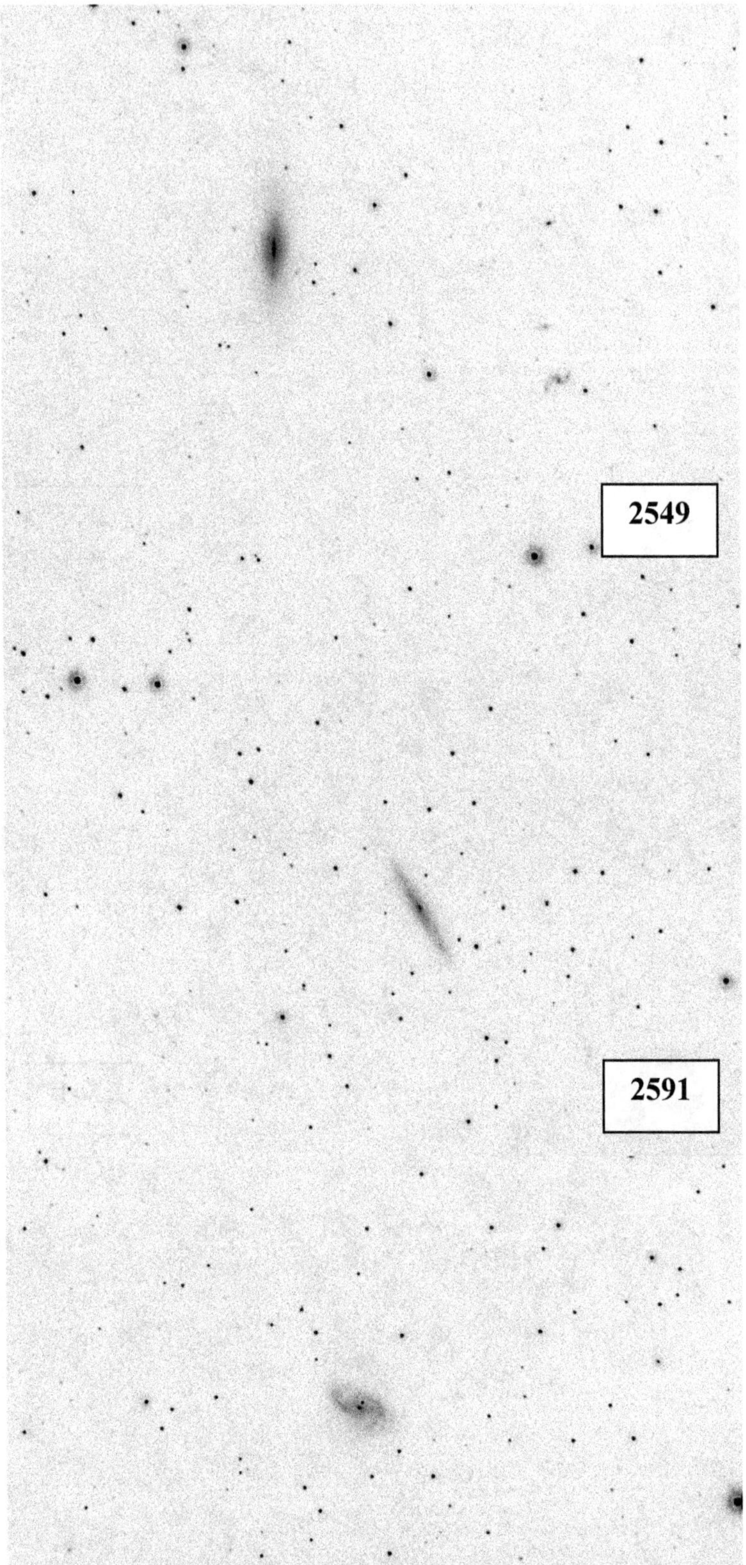

RA 08h

2549

2591

2608

2633

2648

2654

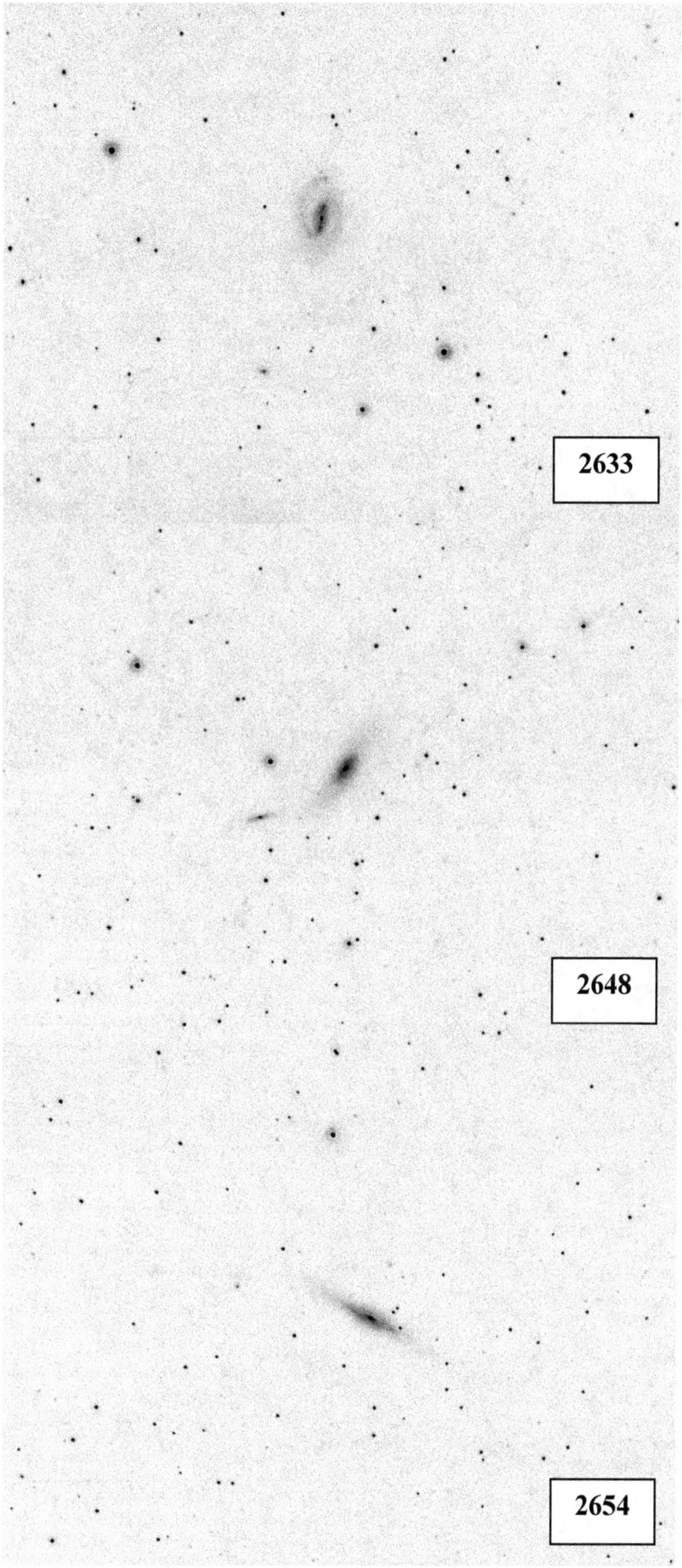

RA 08h

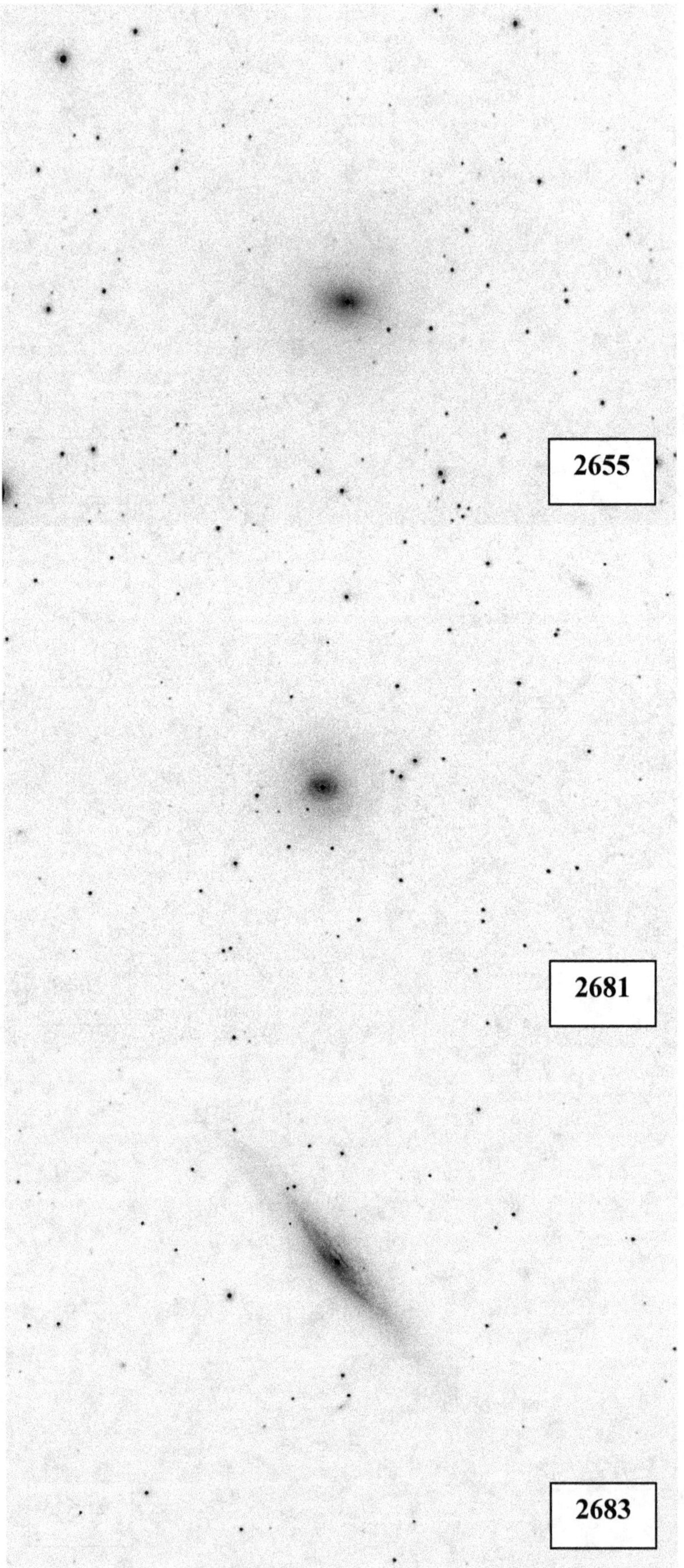

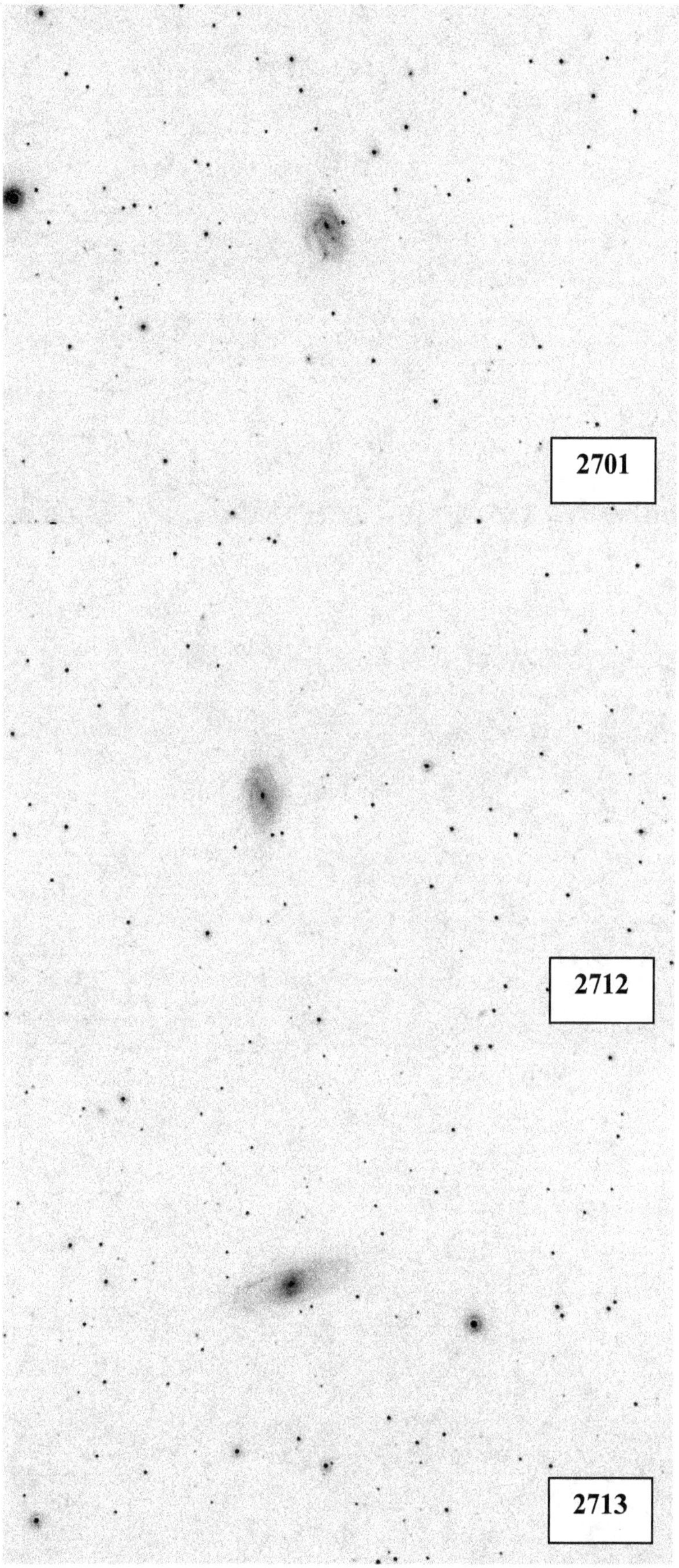

RA 08h

2701

2712

2713

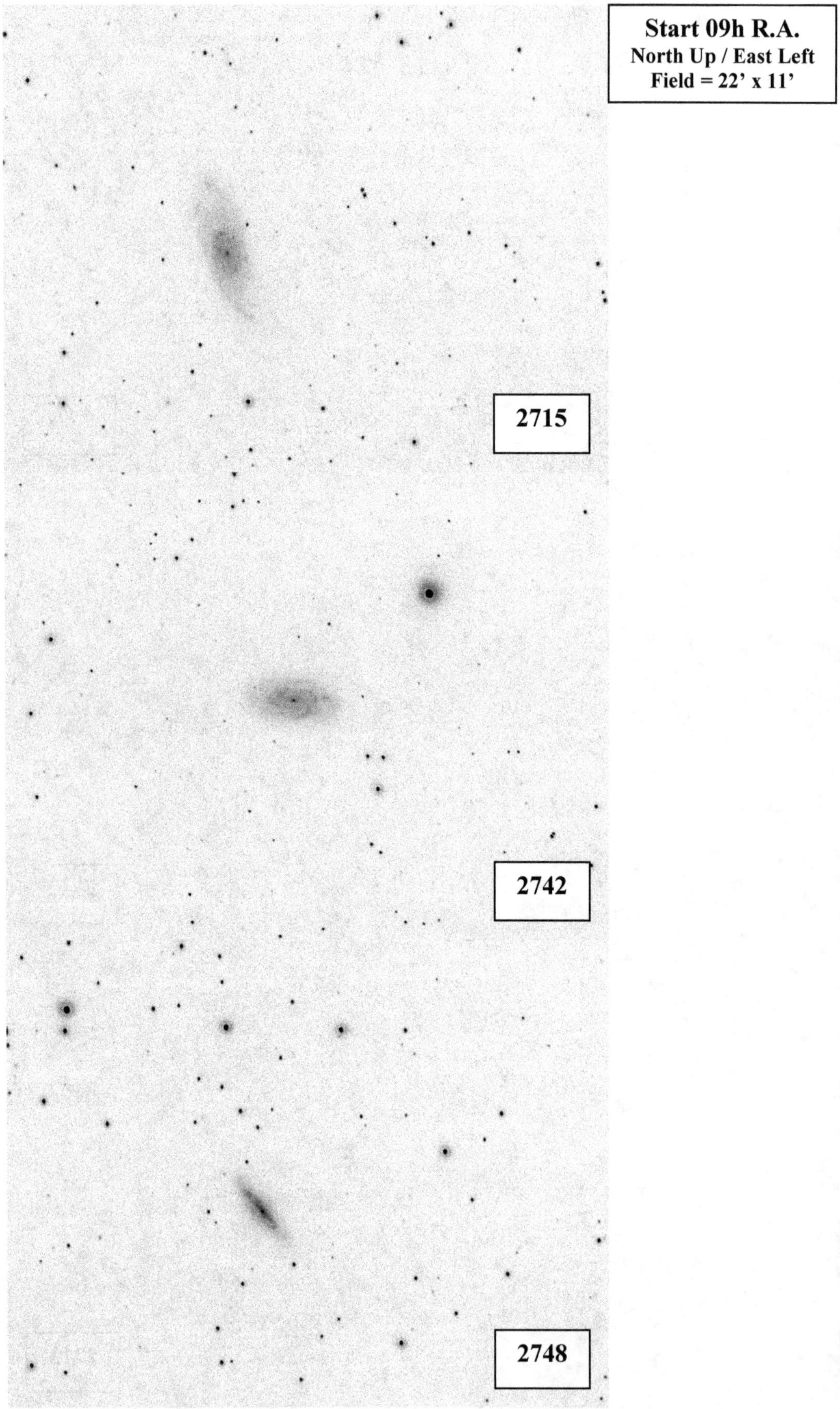

Start 09h R.A.
North Up / East Left
Field = 22' x 11'

2715

2742

2748

RA 09h

2770

2776

2782

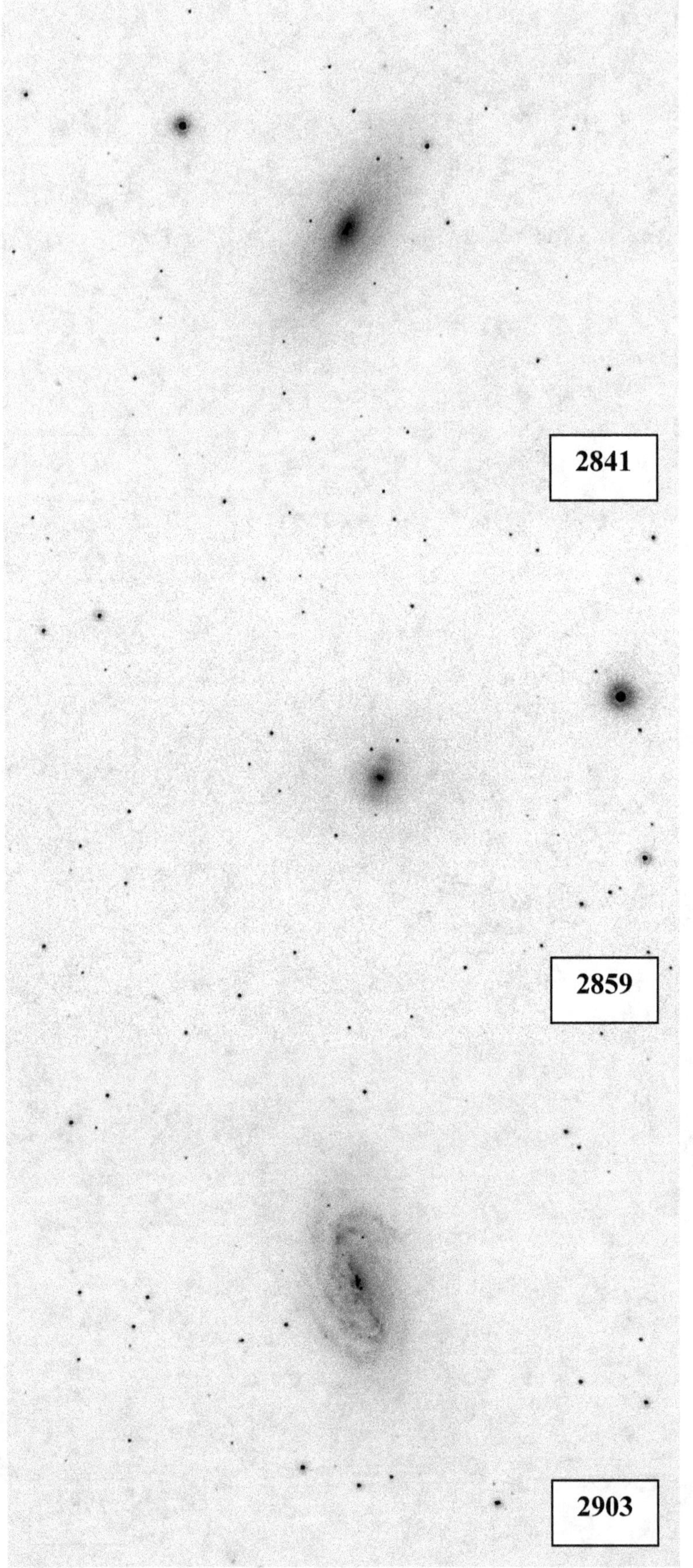

RA 09h

RA 09h

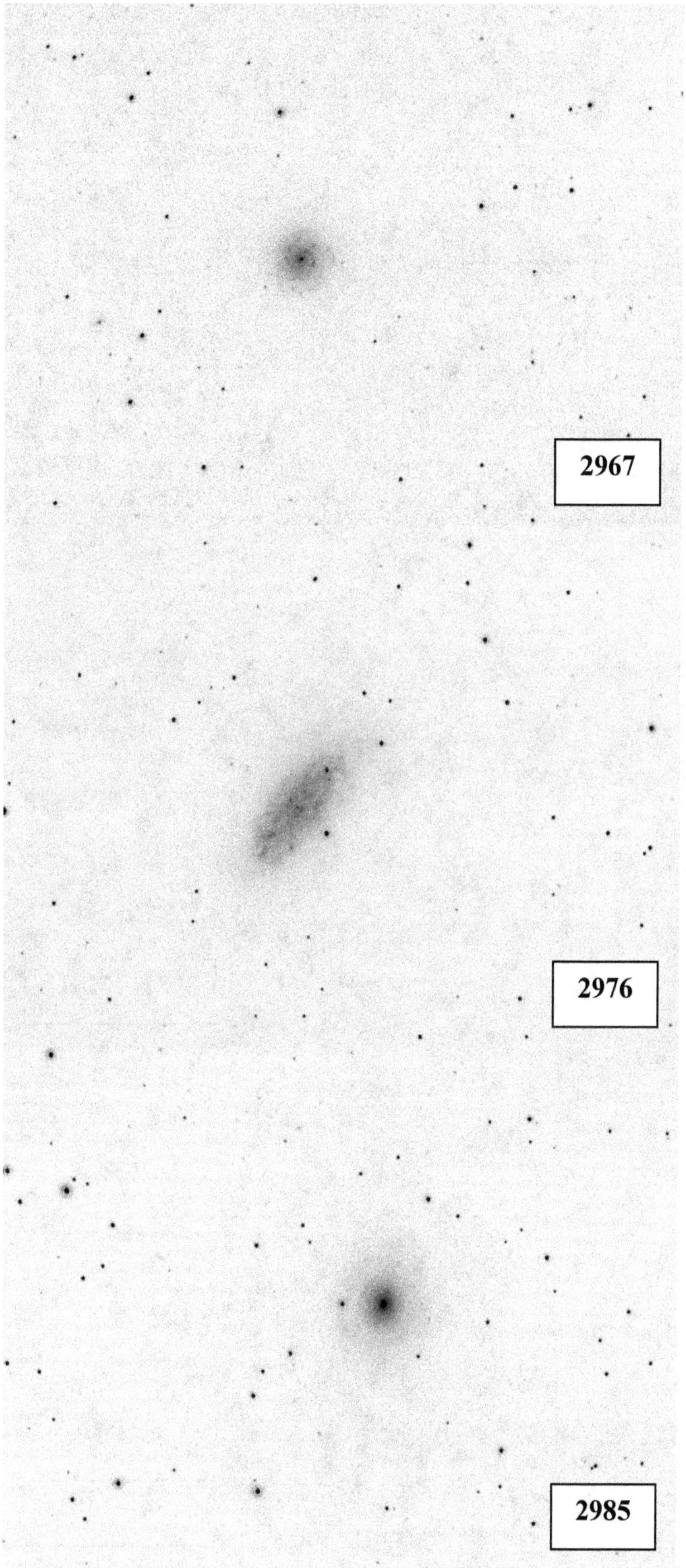

2967

2976

2985

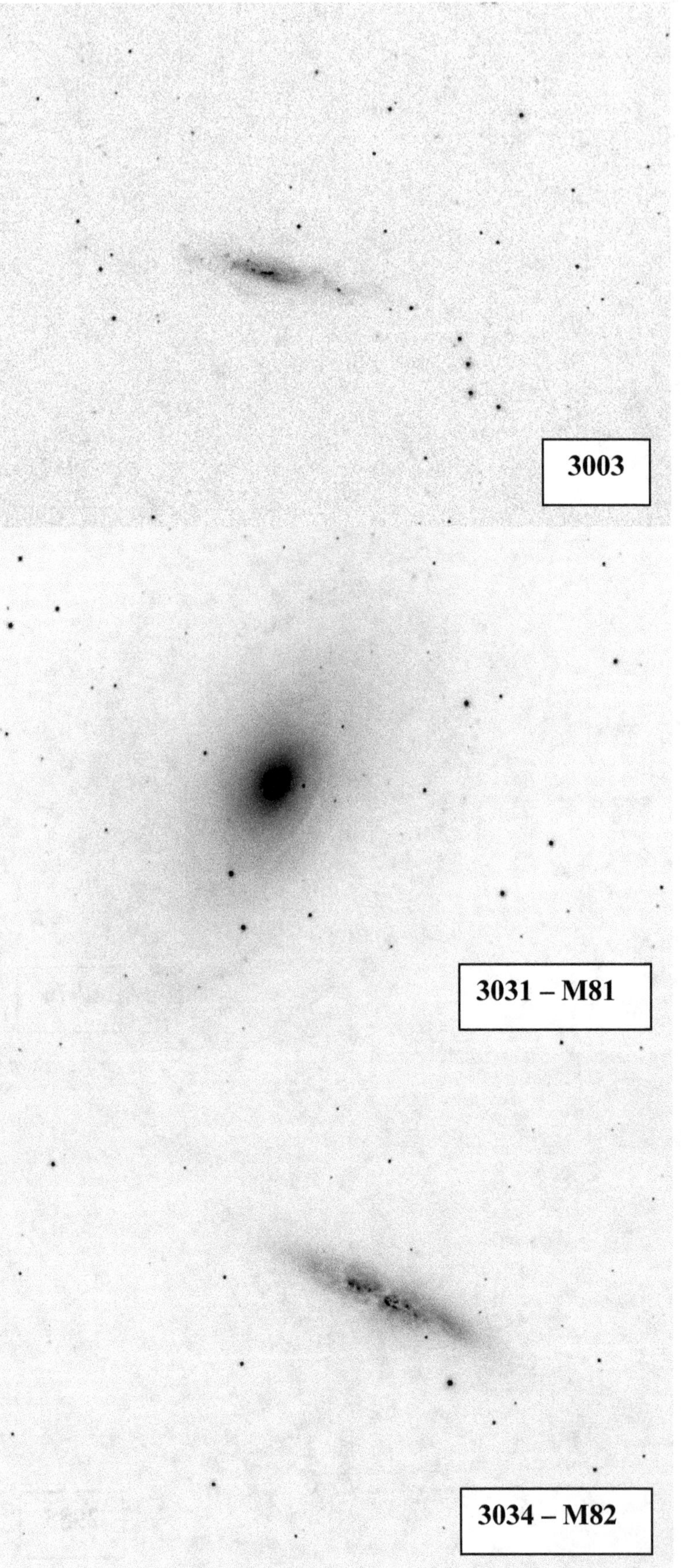
3003
3031 – M81
3034 – M82

RA 09h

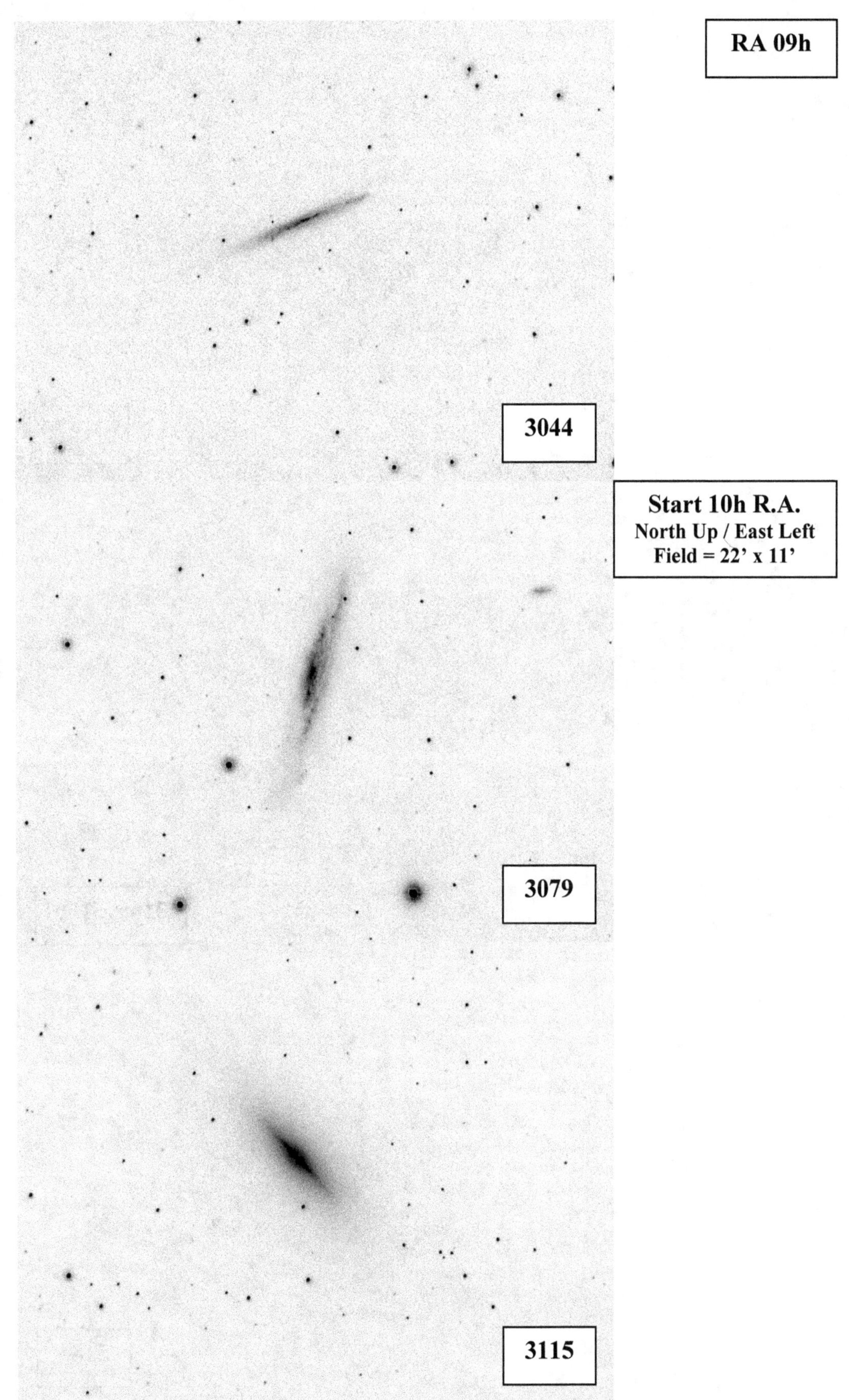

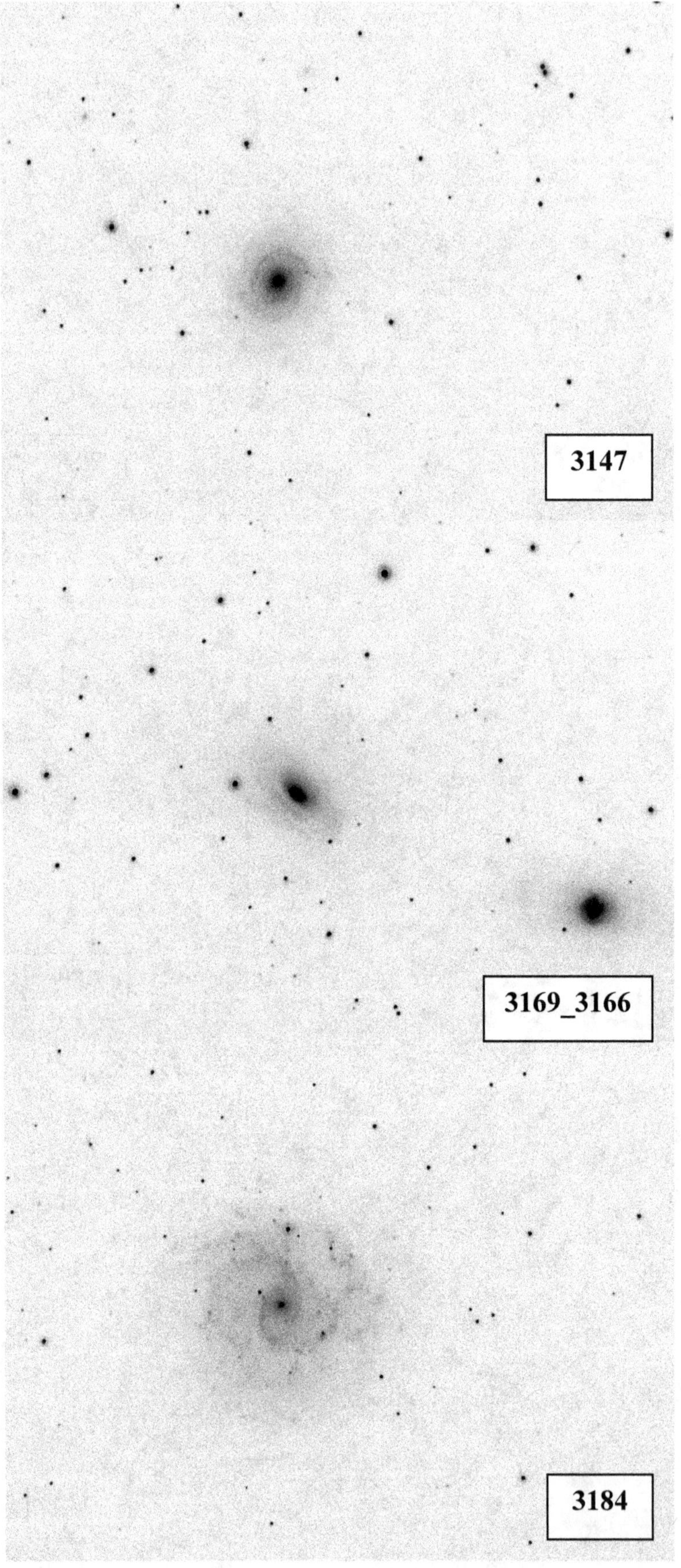
3147
3169_3166
3184

RA 10h

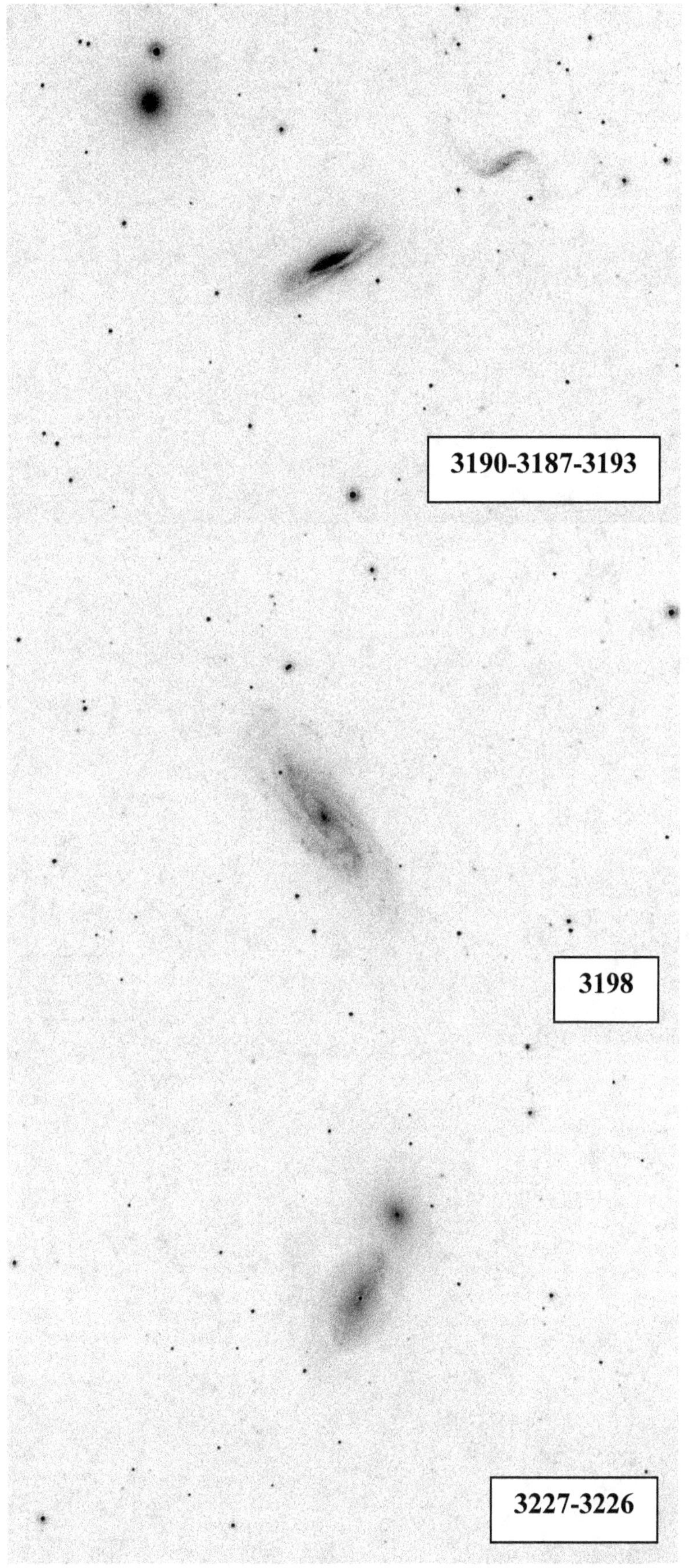

3254

3294

3338

RA 10h

3344

3346

3351 – M95

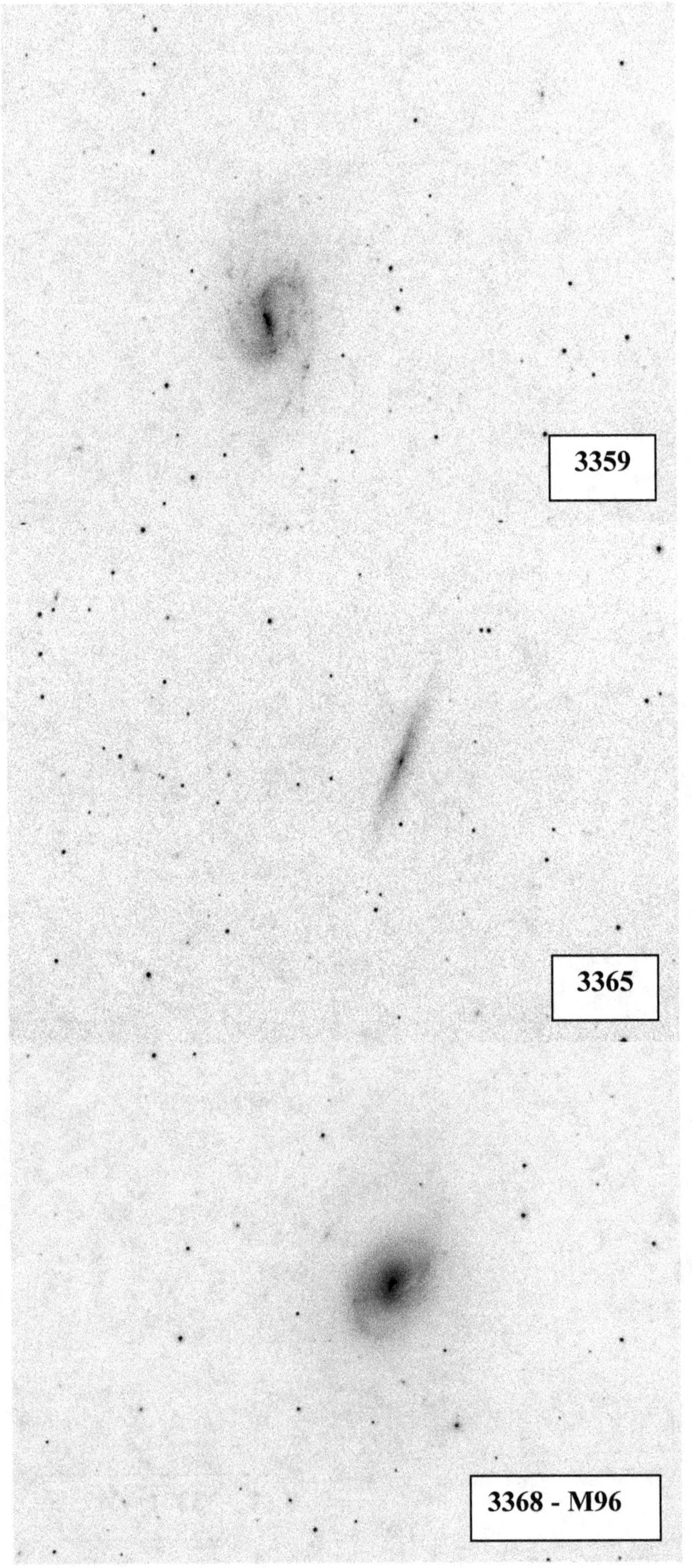
3359
3365
3368 - M96

RA 10h

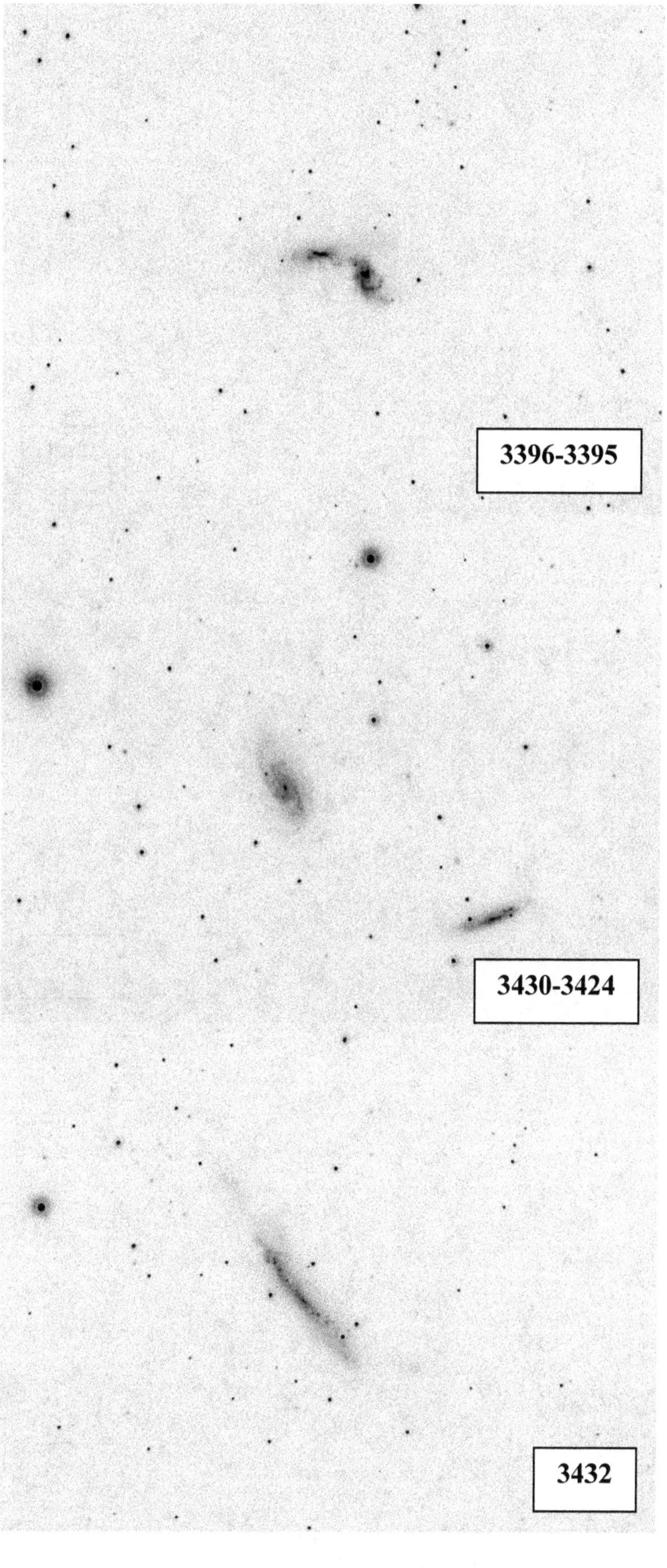

3396-3395

3430-3424

3432

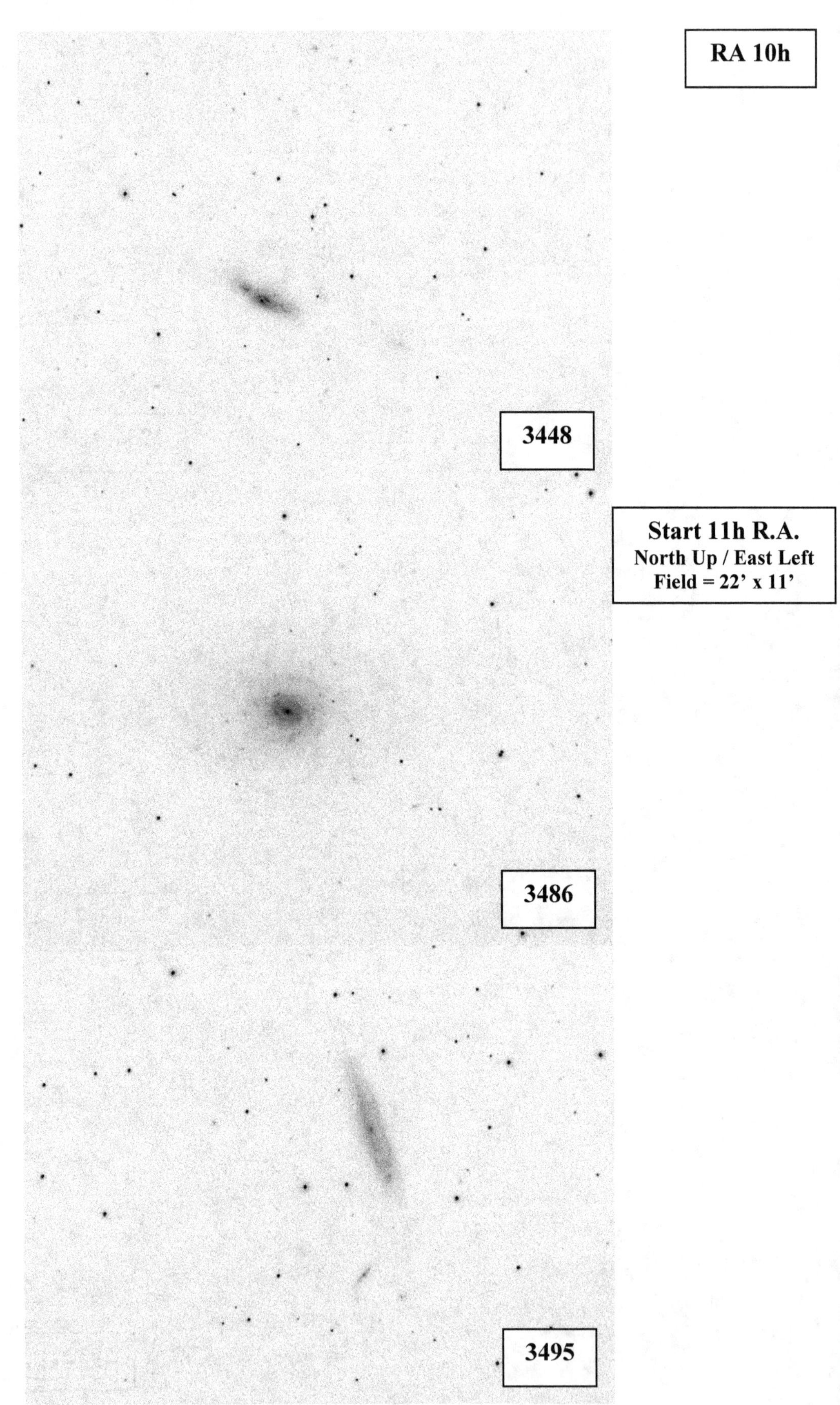
RA 10h
3448
Start 11h R.A.
North Up / East Left
Field = 22' x 11'
3486
3495

RA 11h

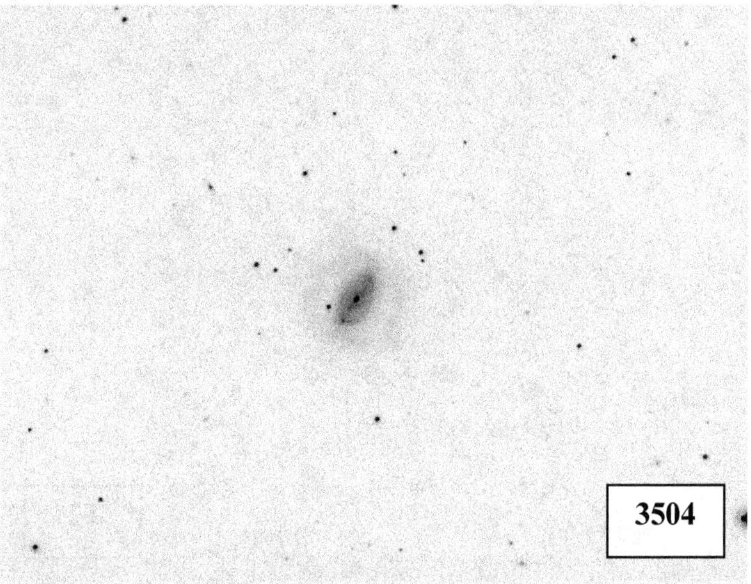

3504

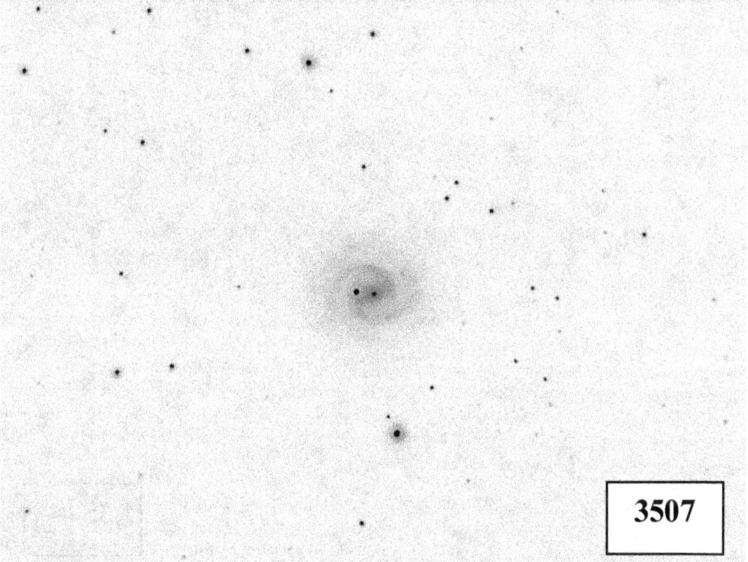

3507

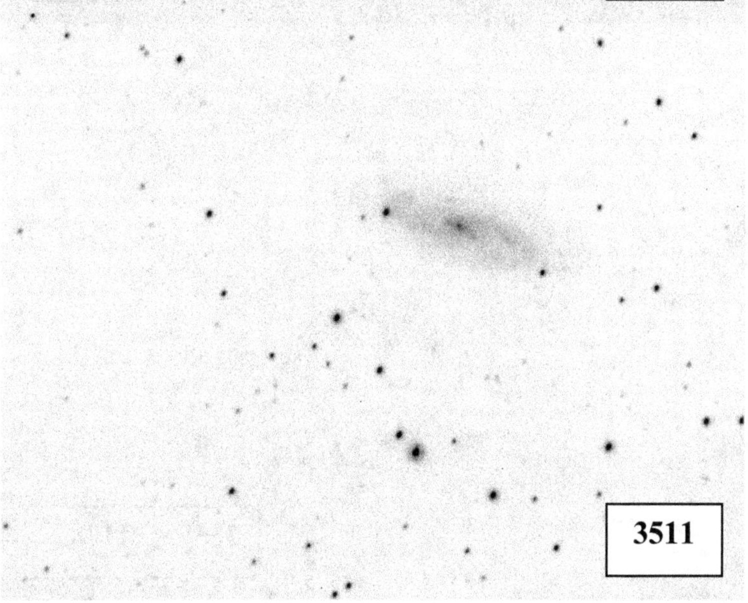

3511

3521

3549

3556 – M108

RA 11h

3583

3593

3596

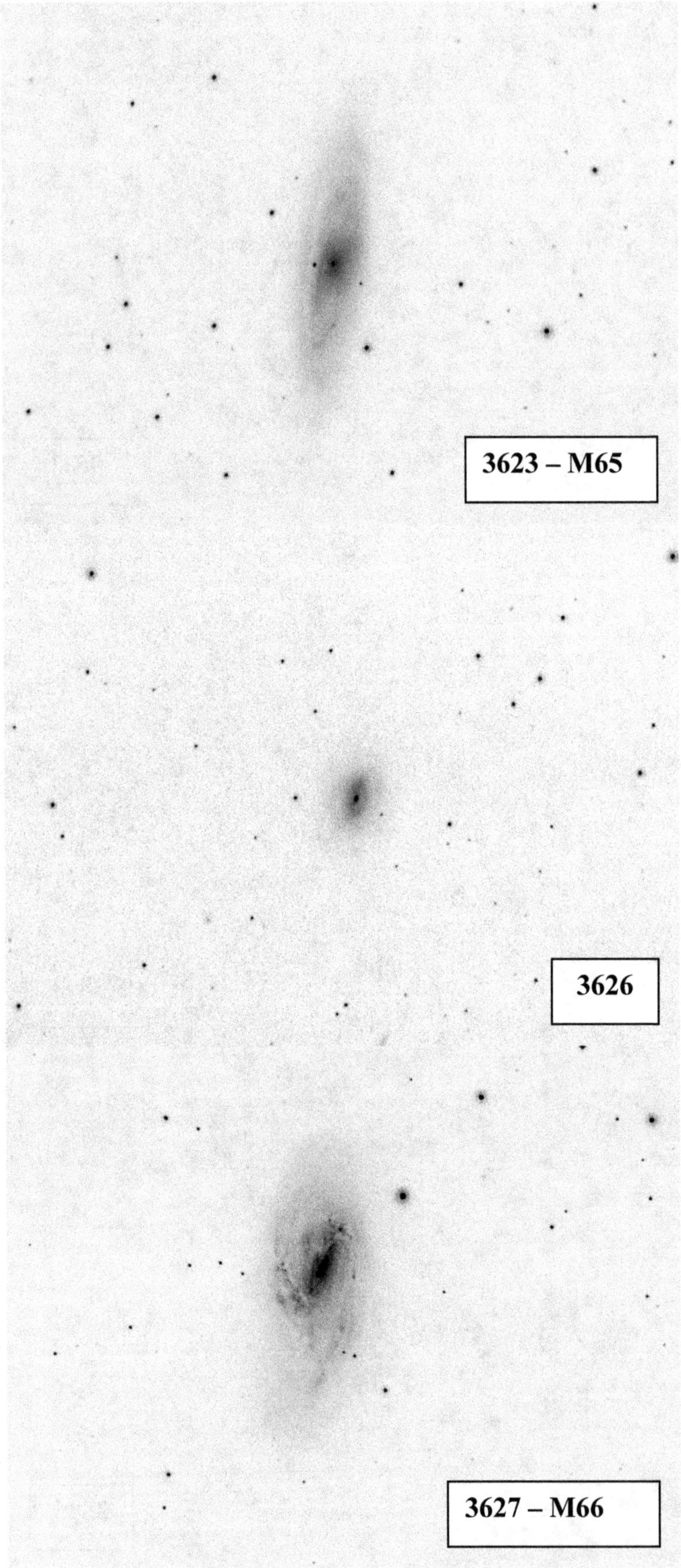

RA 11h

3623 – M65

3626

3627 – M66

RA 11h

3628

3631

3642

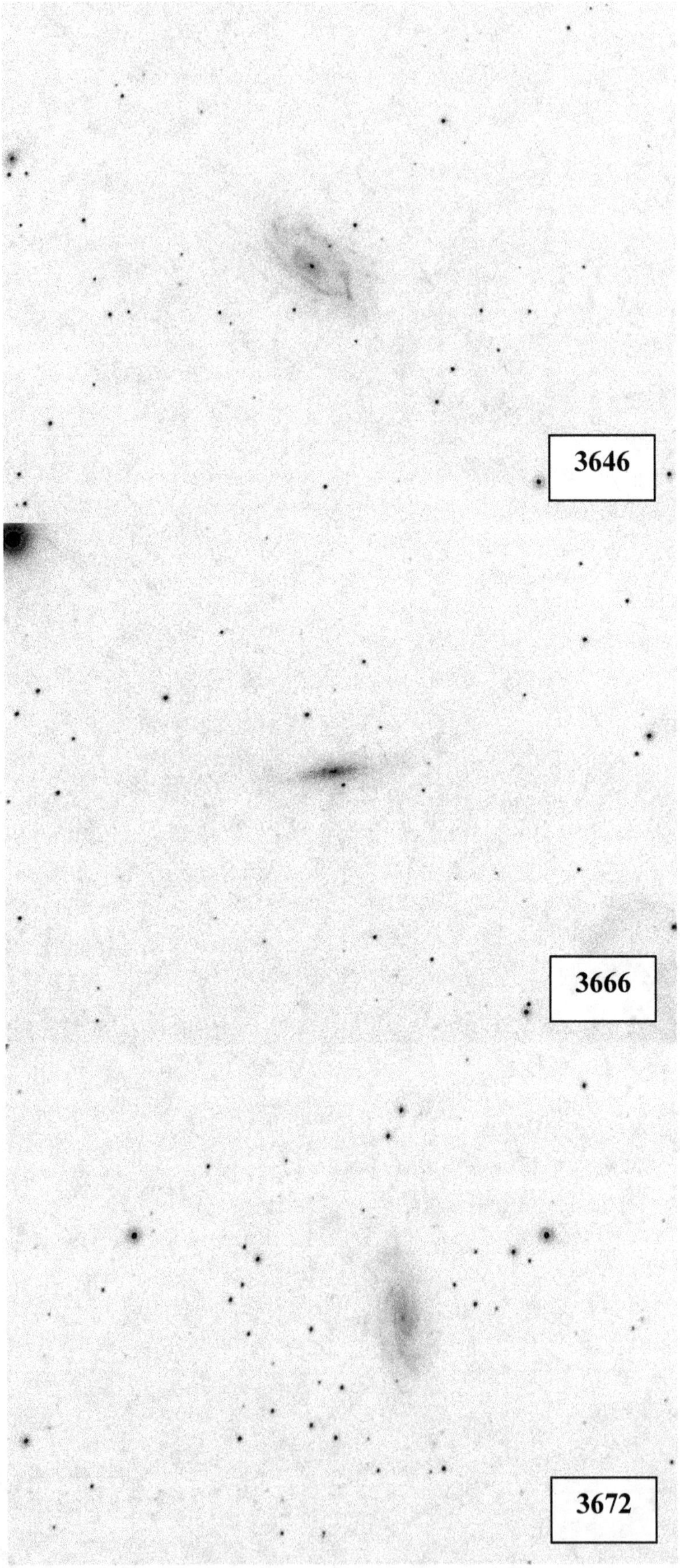

3646

3666

3672

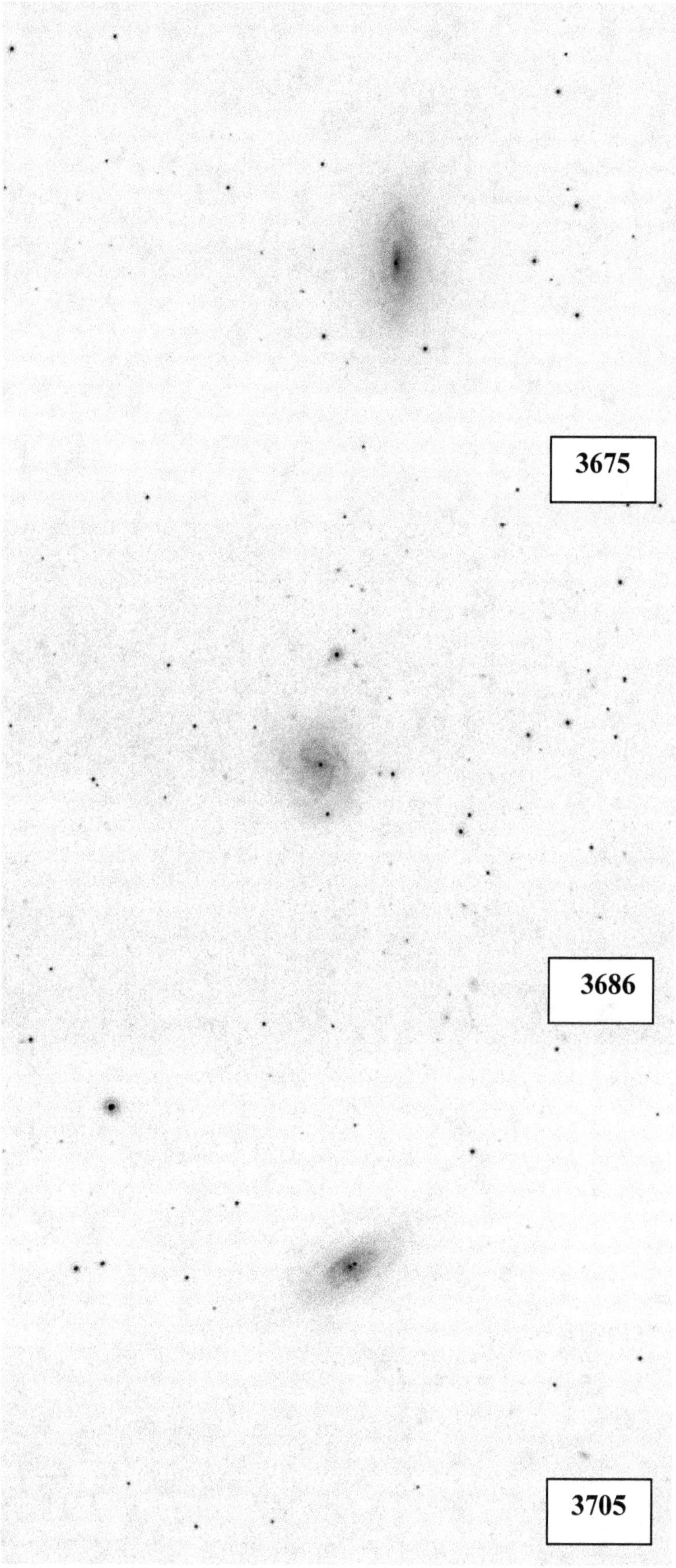

RA 11h

3675

3686

3705

RA 11h

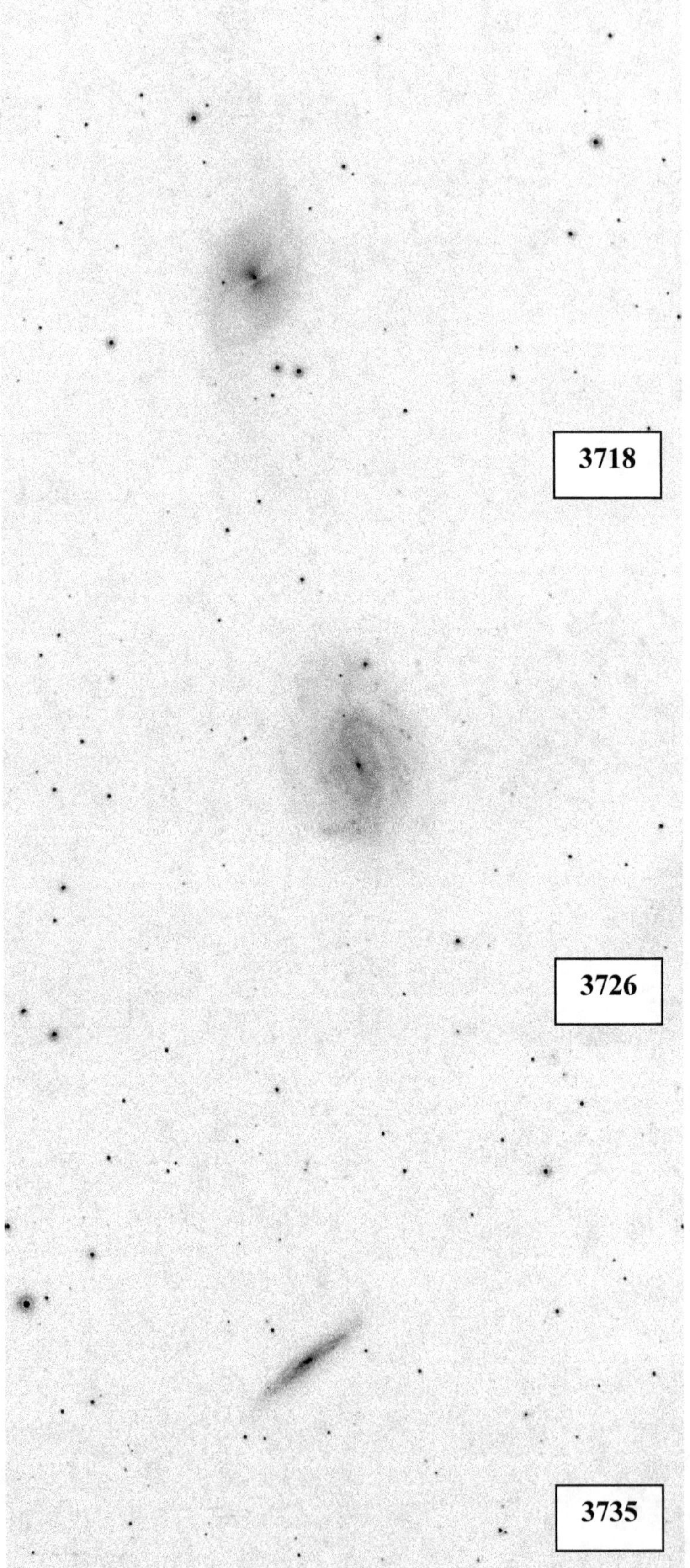

RA 11h

3756

3780

3810

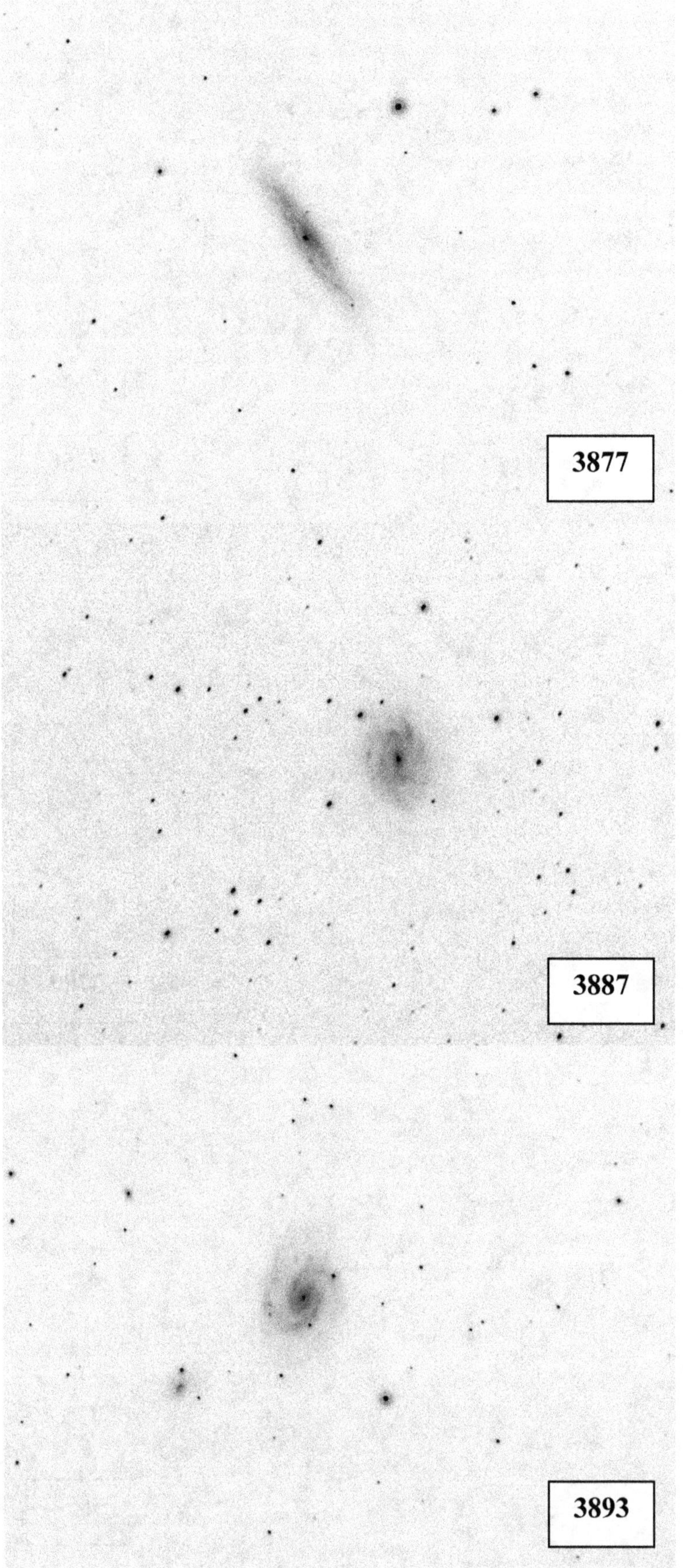
3877
3887
3893

RA 11h

3898

3938

3953

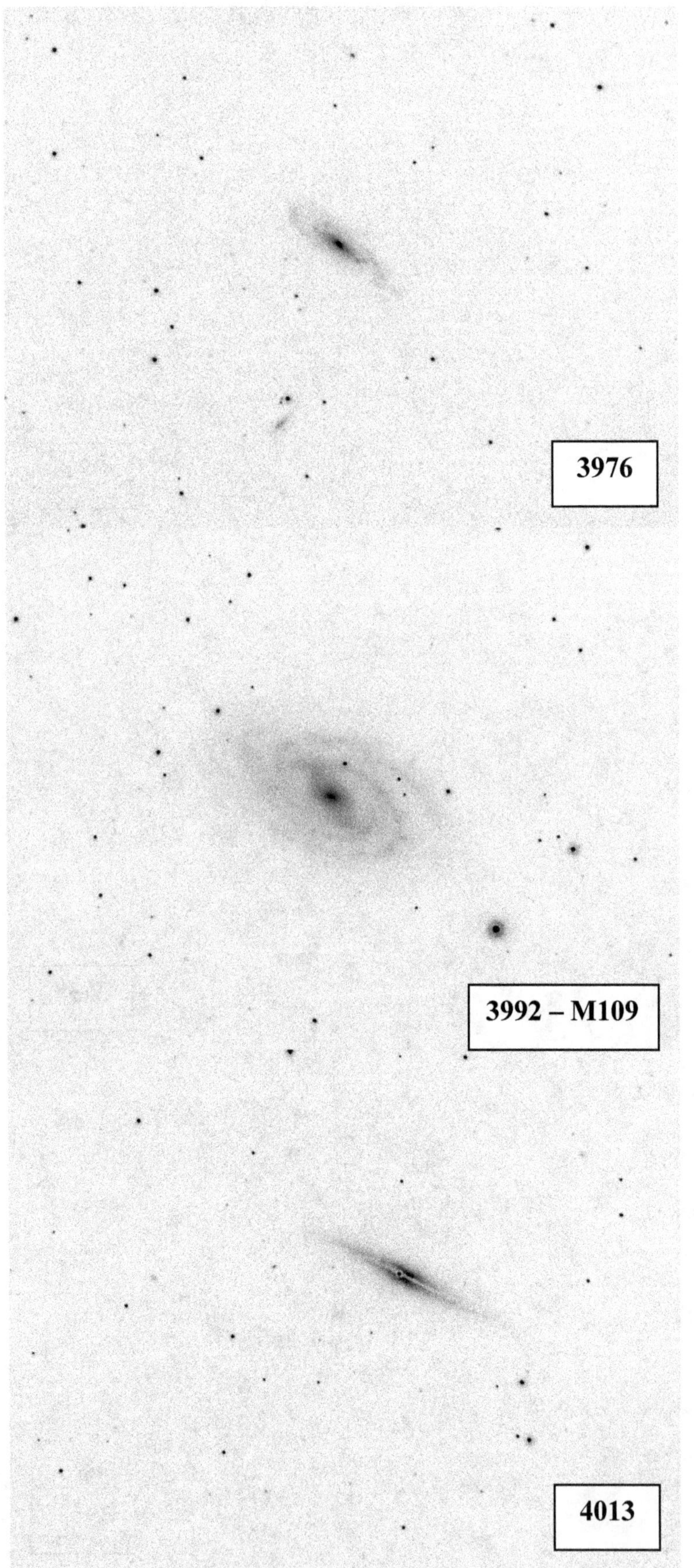
3976
3992 – M109
4013

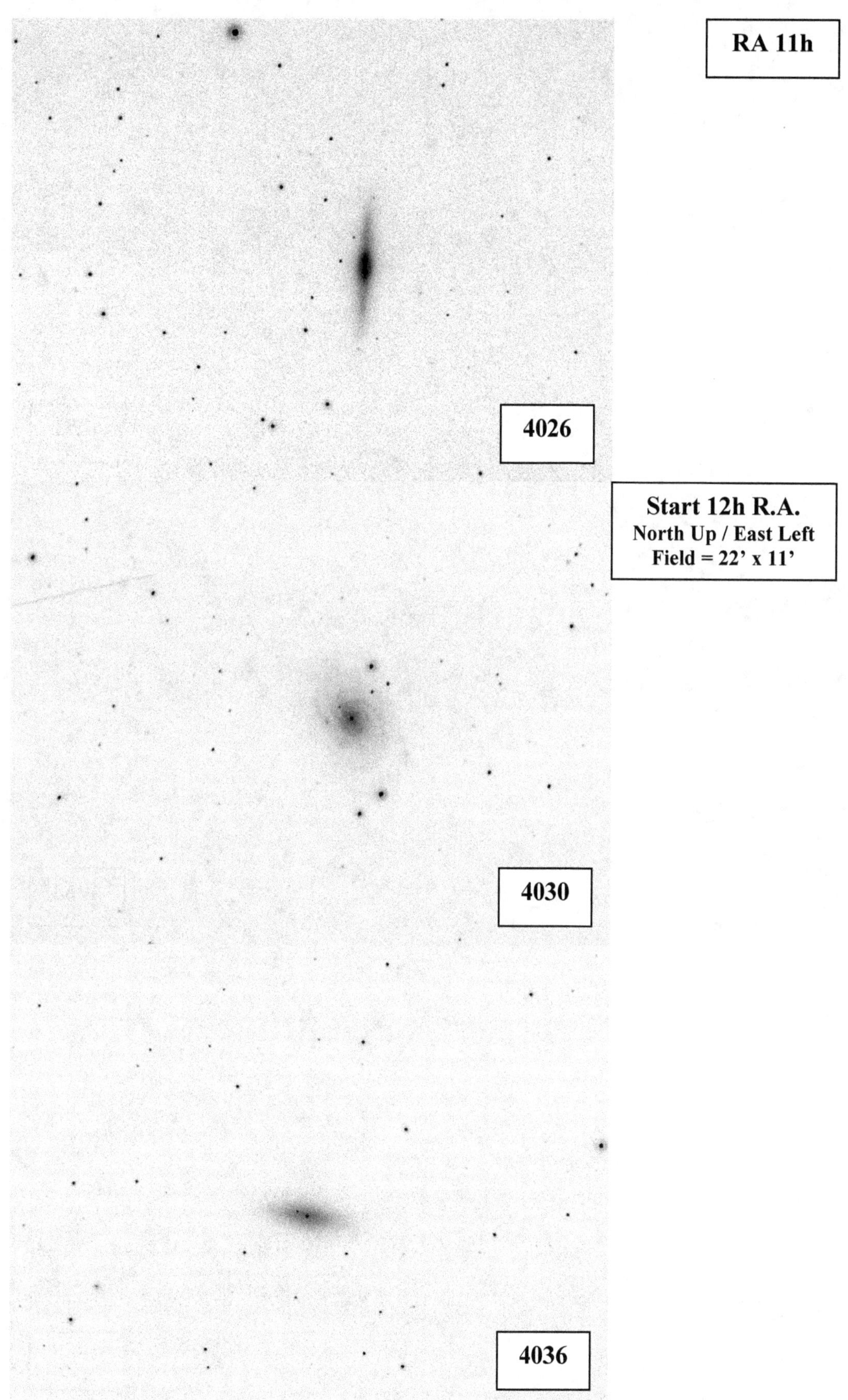
RA 11h
4026
Start 12h R.A.
North Up / East Left
Field = 22' x 11'
4030
4036

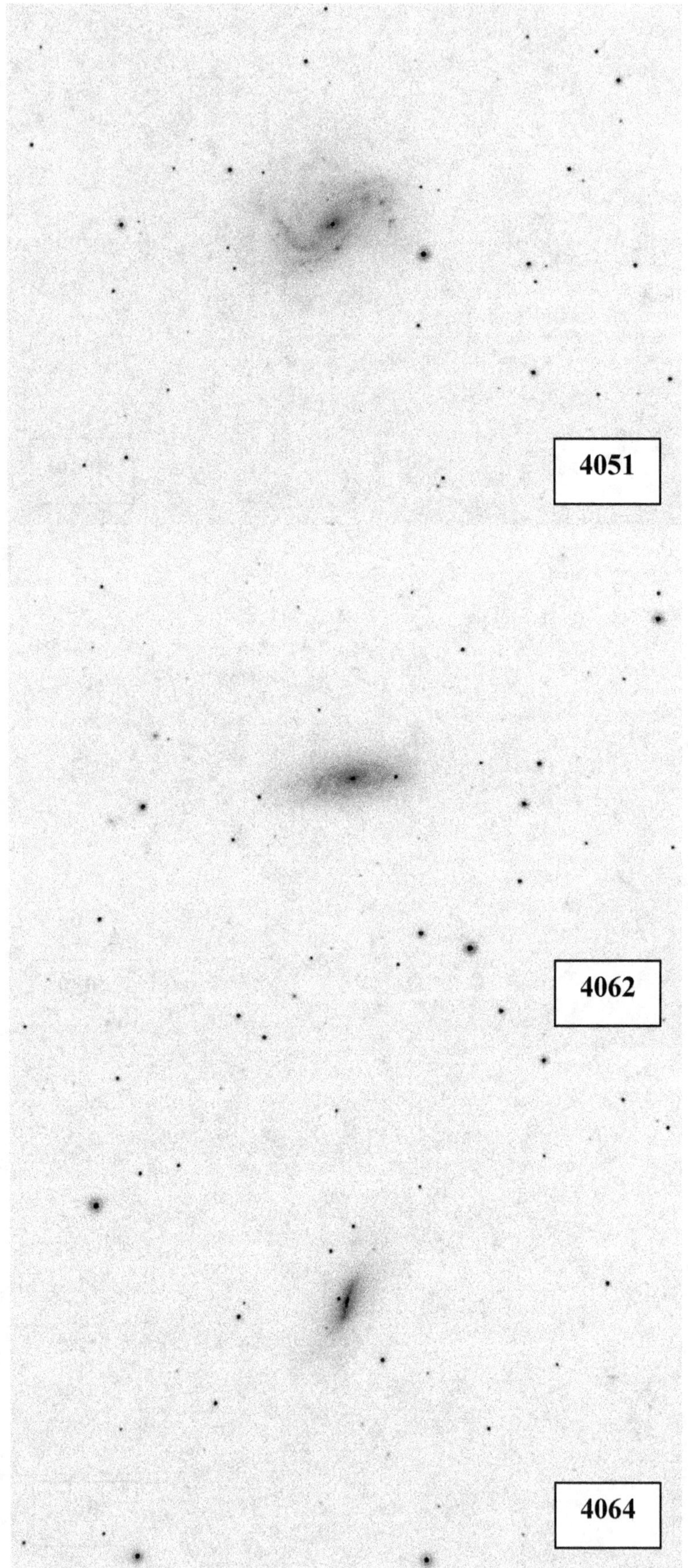
4051
4062
4064

RA 12h

4088

4100

4111

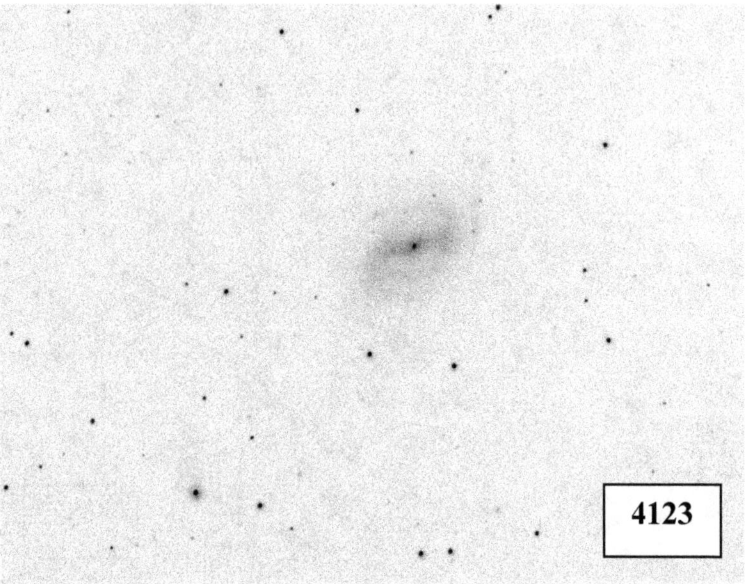

4123

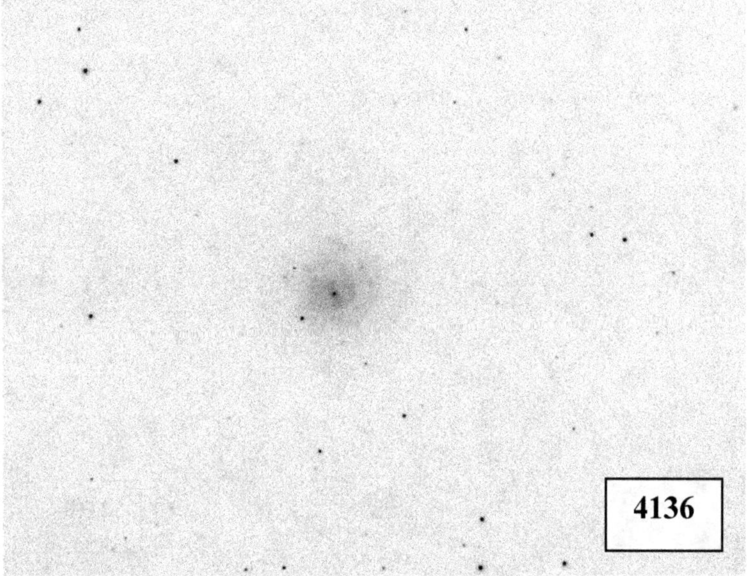

4136

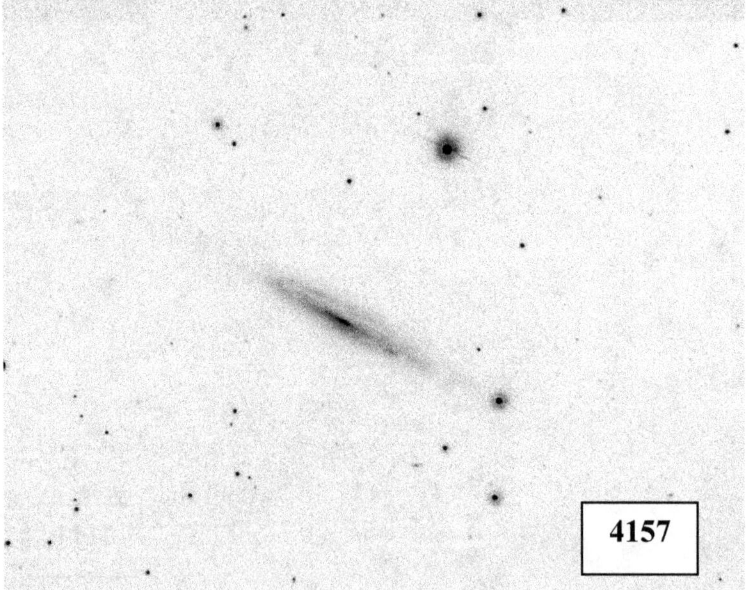

4157

RA 12h

4179

4192 – M98

4214

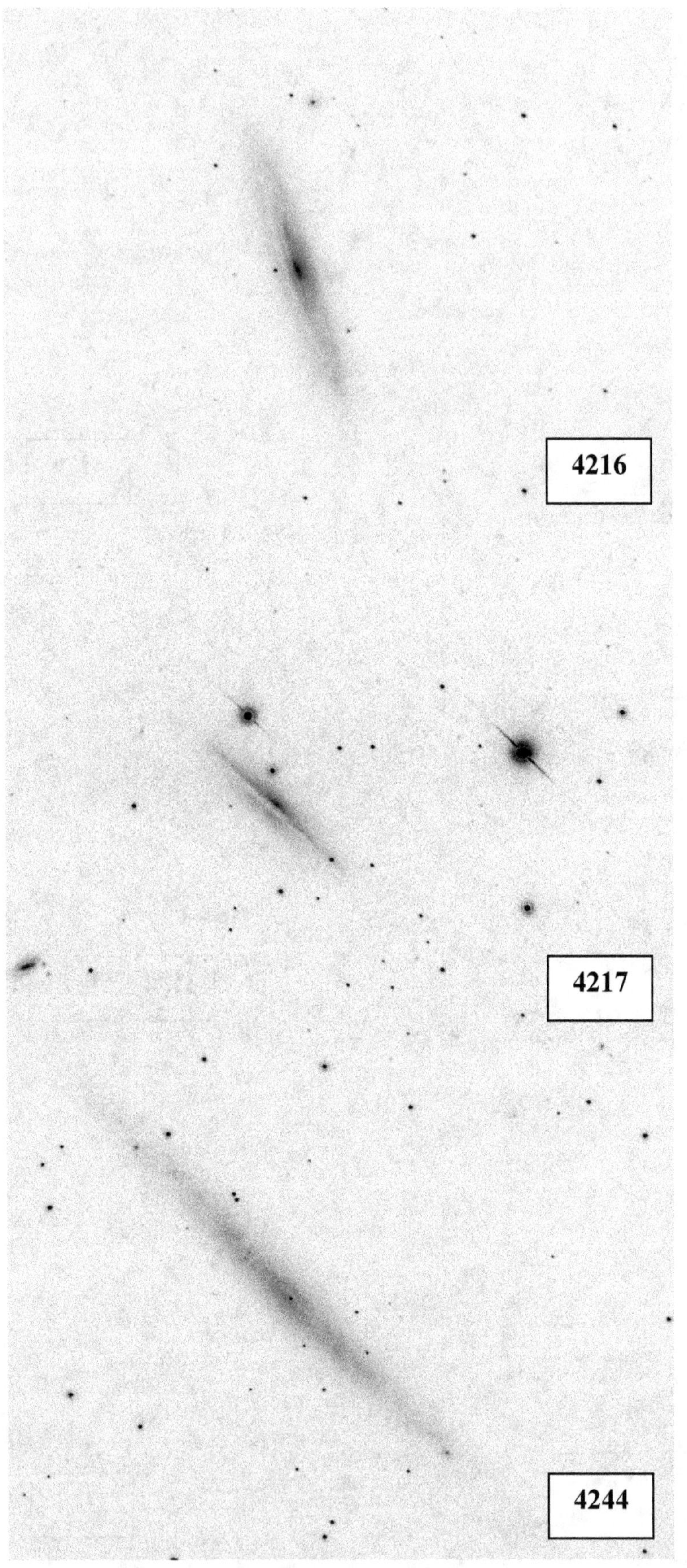
4216
4217
4244

RA 12h

4254 – M99

4258 – M106

4274

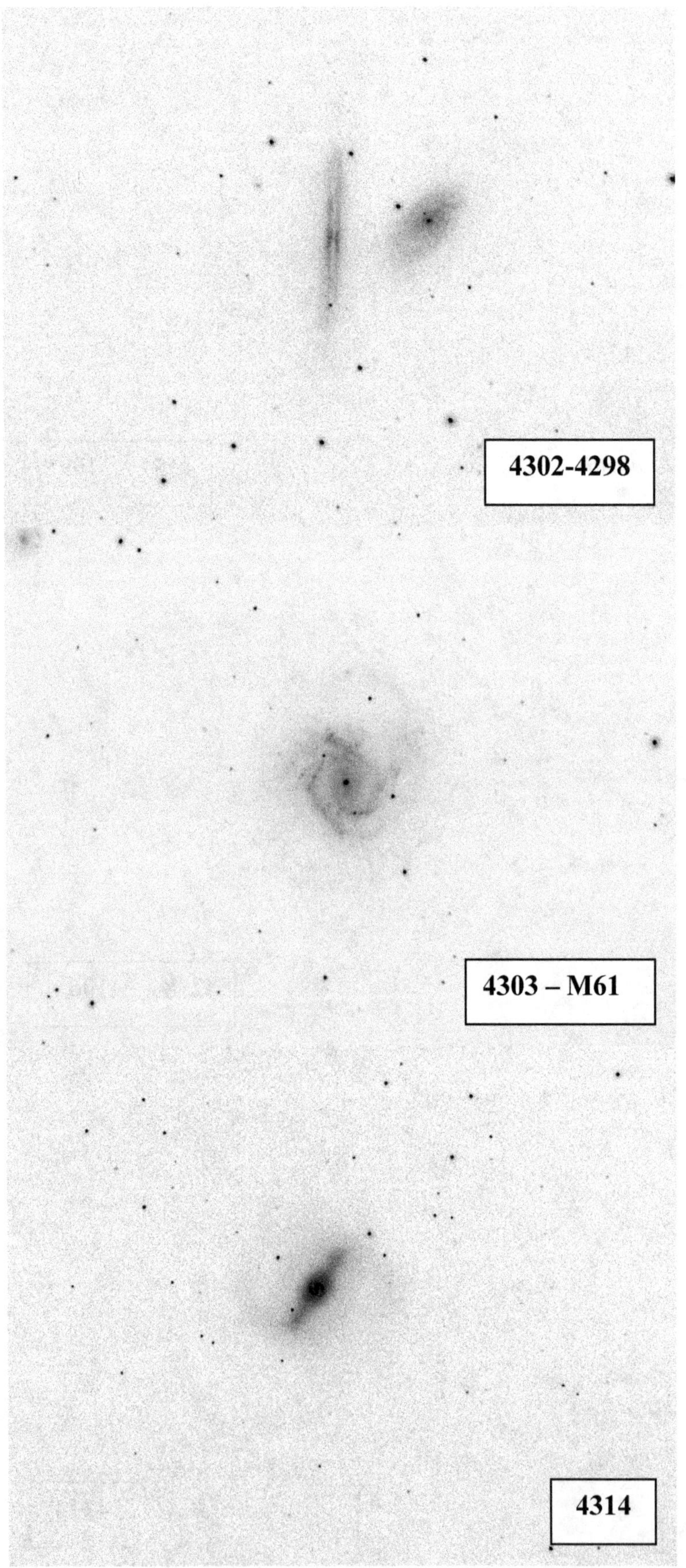
4302-4298
4303 – M61
4314

4321 – M100

4395

4414

RA 12h

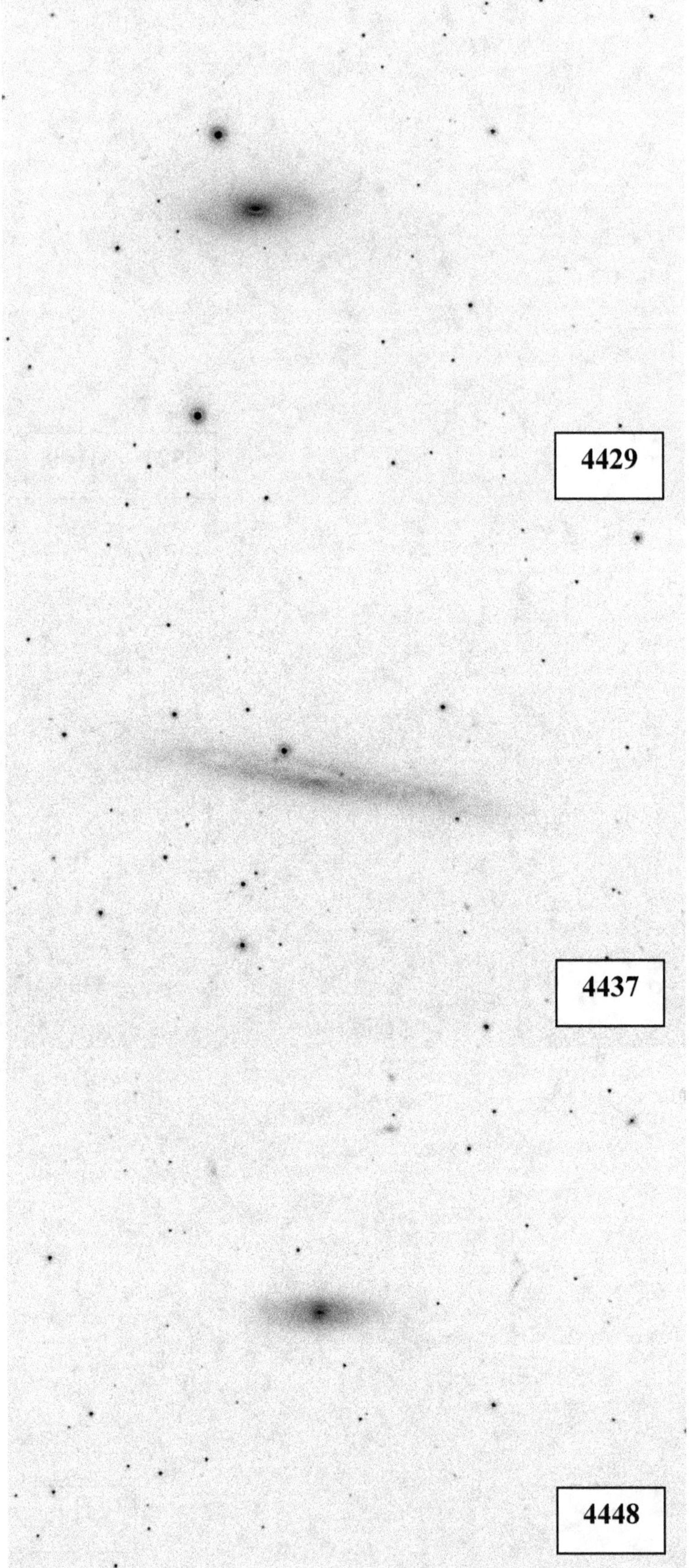

RA 12h

4449

4450

4490

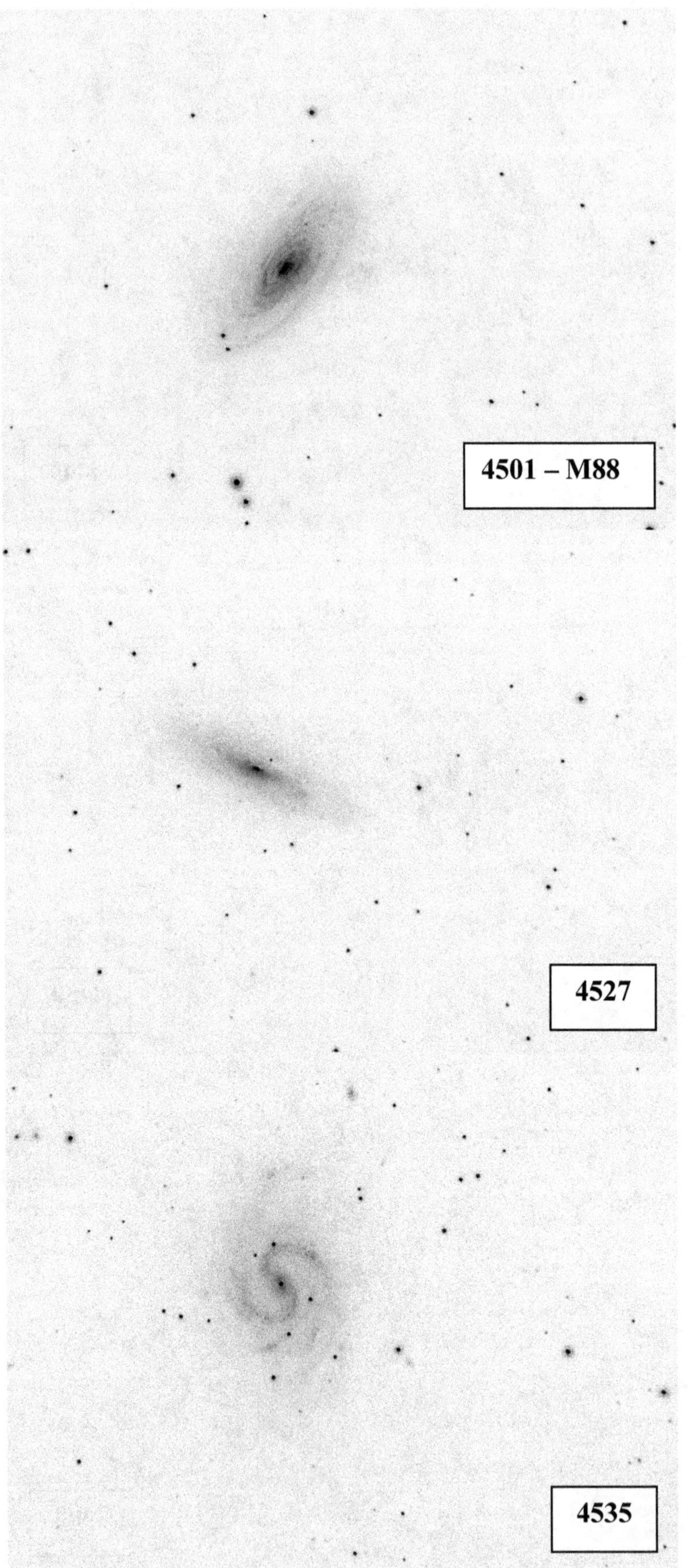
4501 – M88
4527
4535

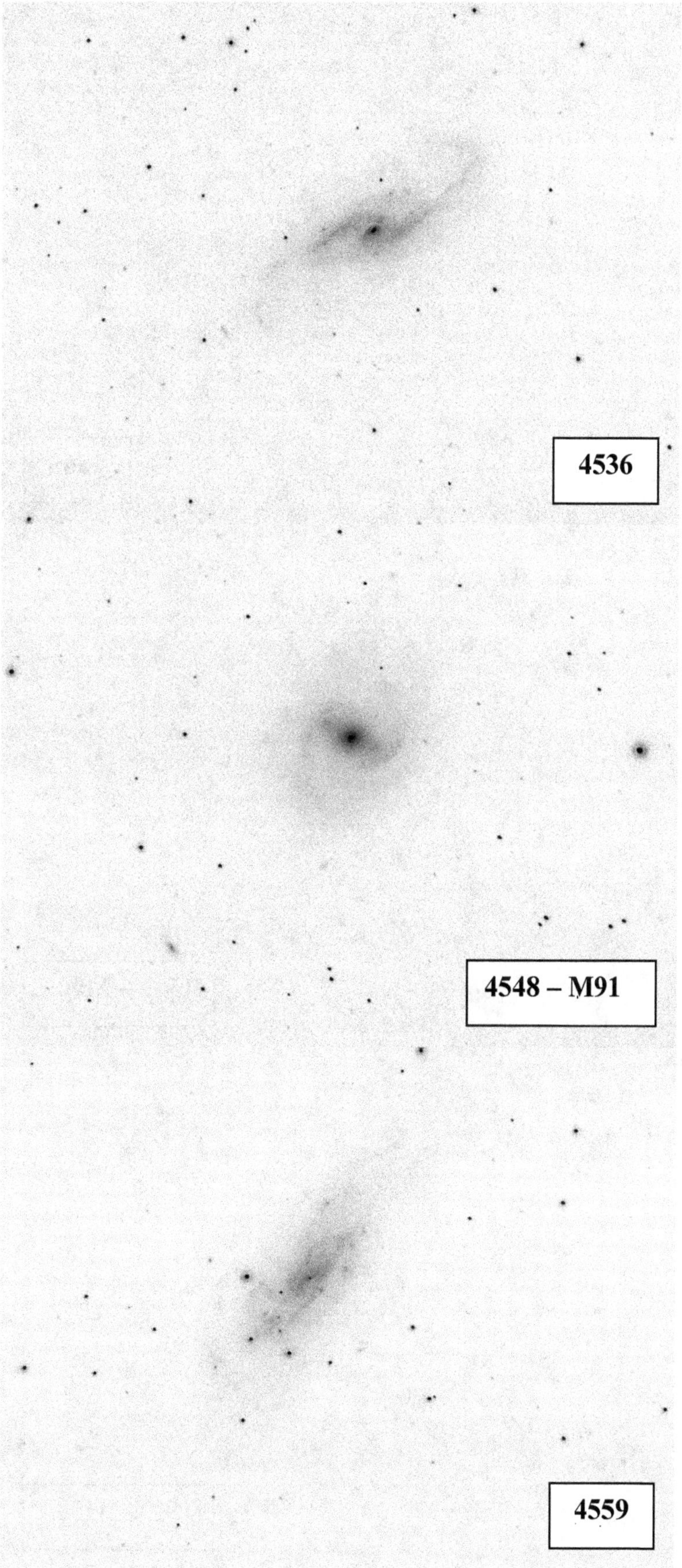

RA 12h

4536

4548 – M91

4559

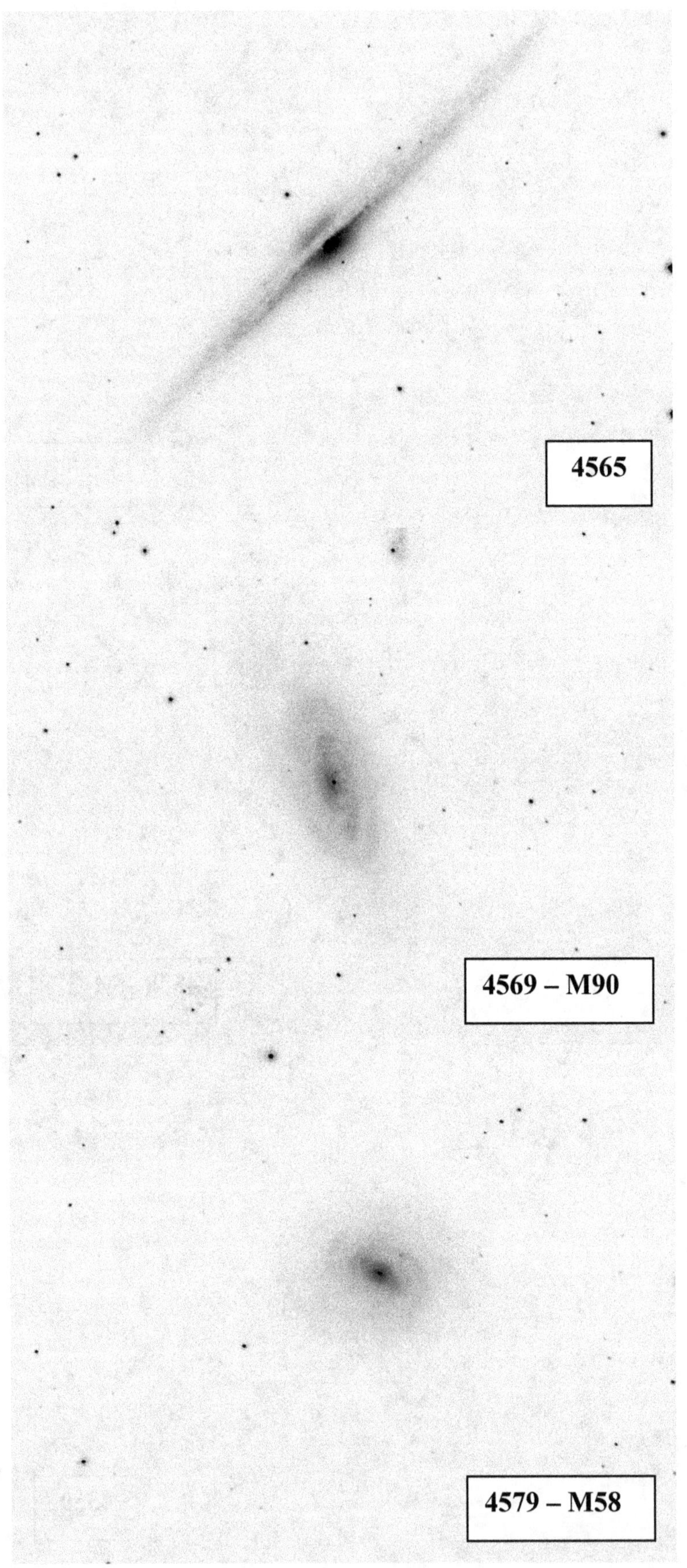
4565
4569 – M90
4579 – M58

4593

4594 – M104

4605

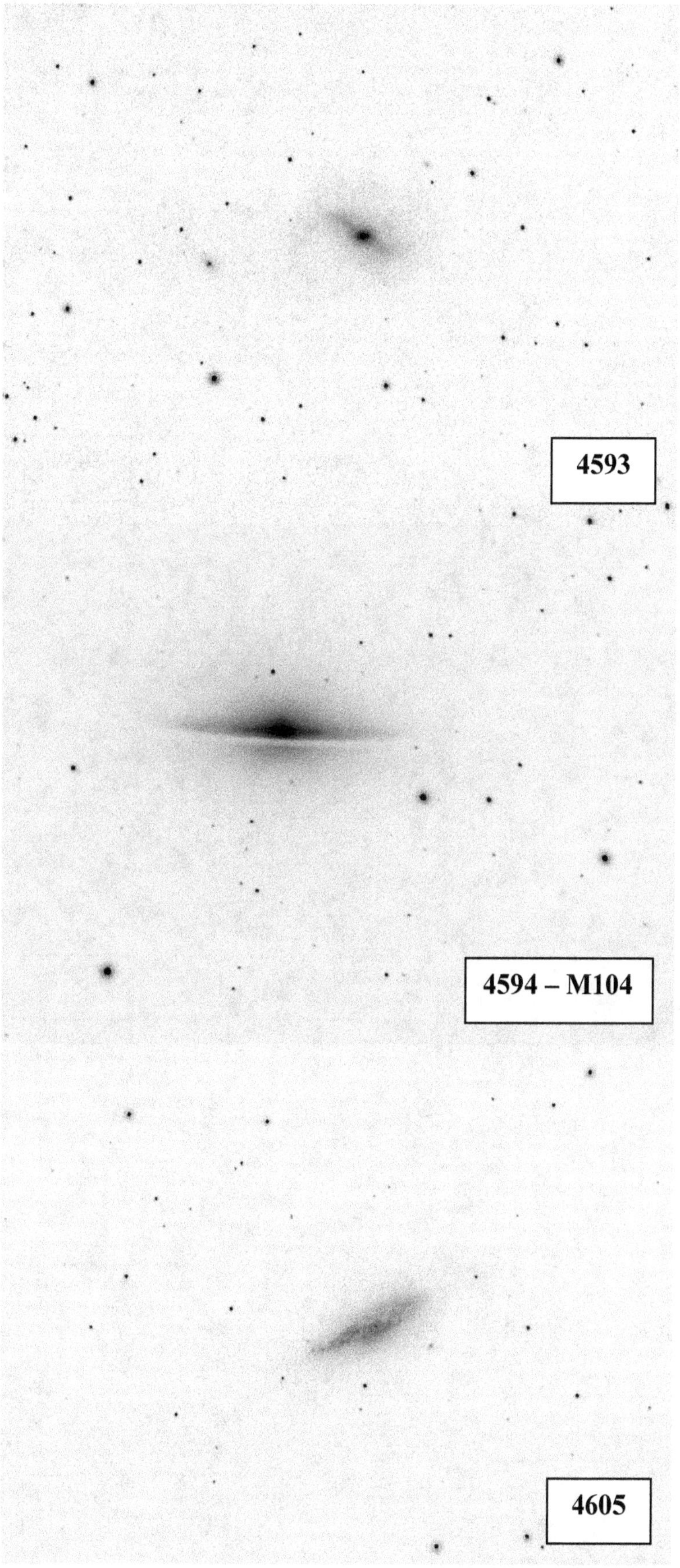

RA 12h

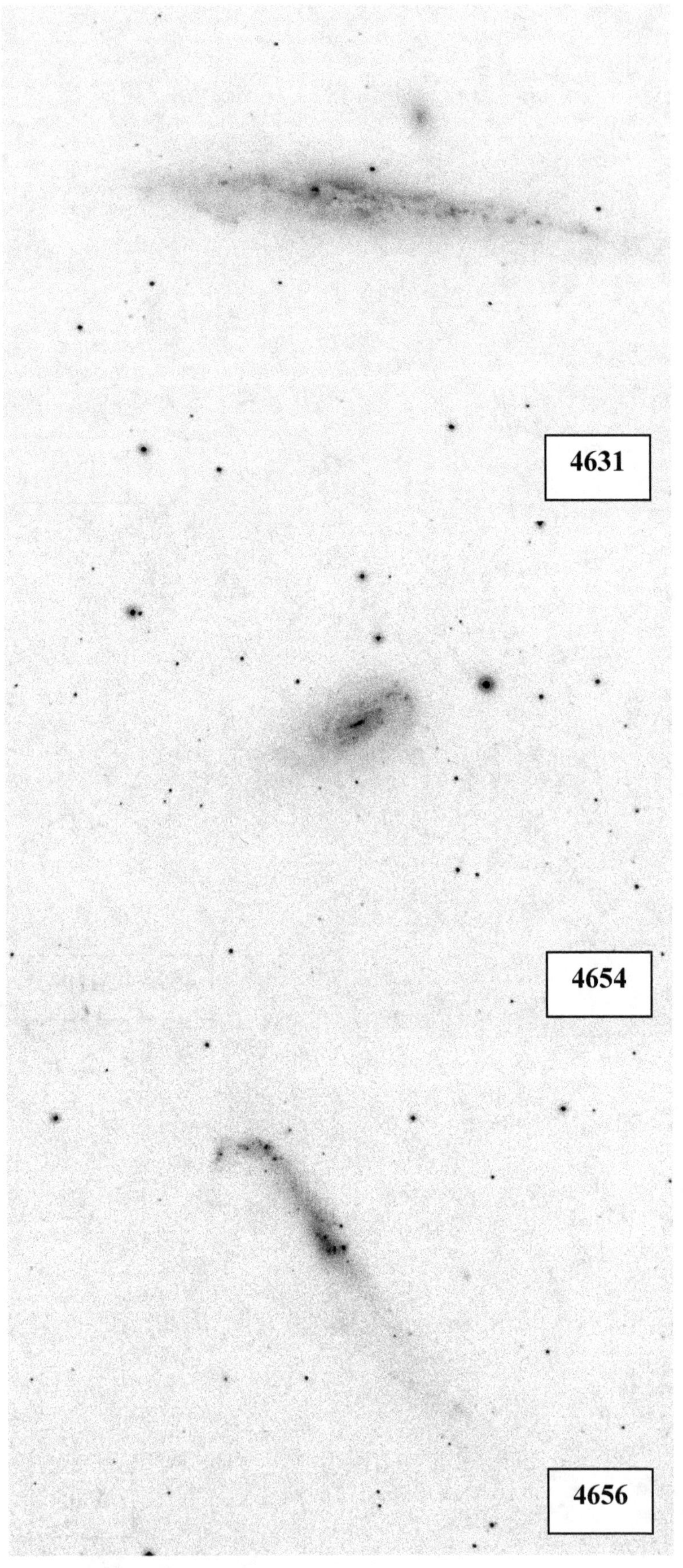

RA 12h

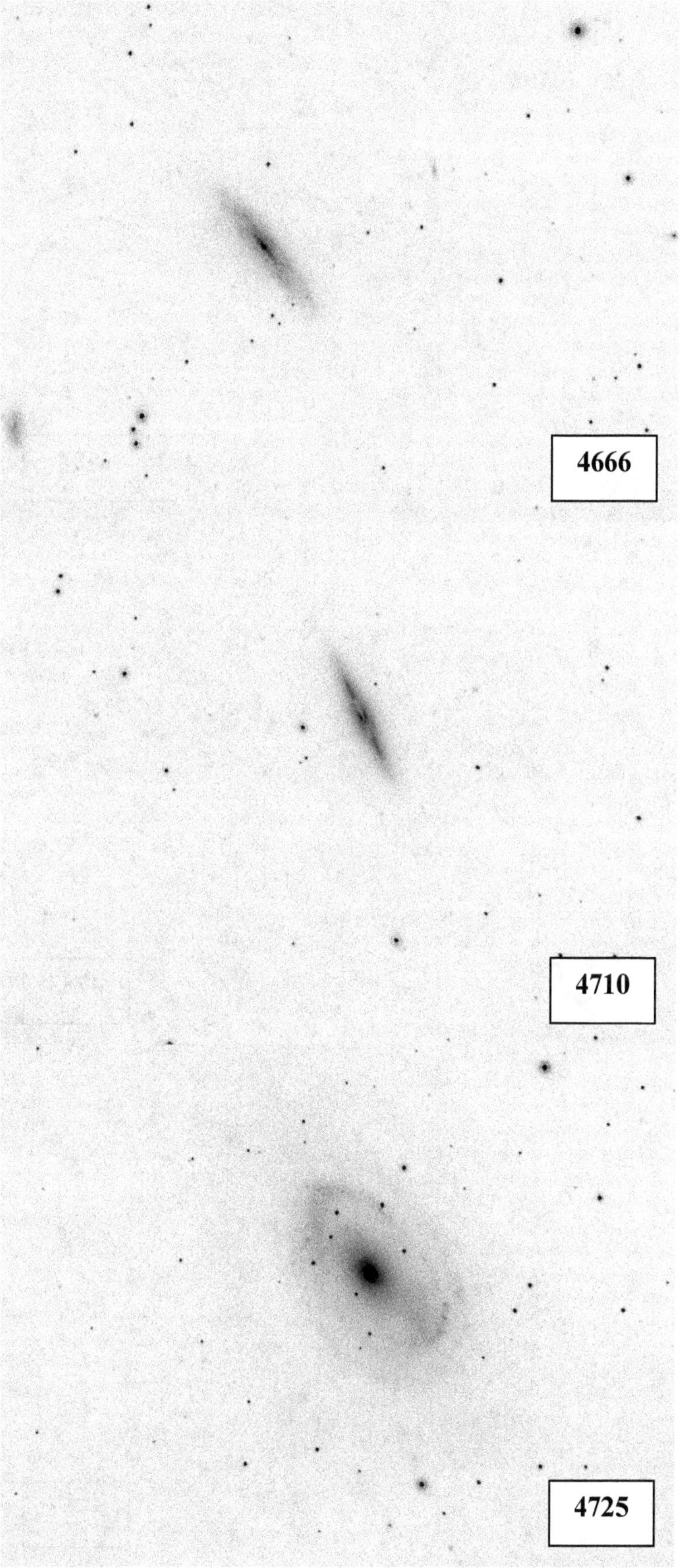

4666

4710

4725

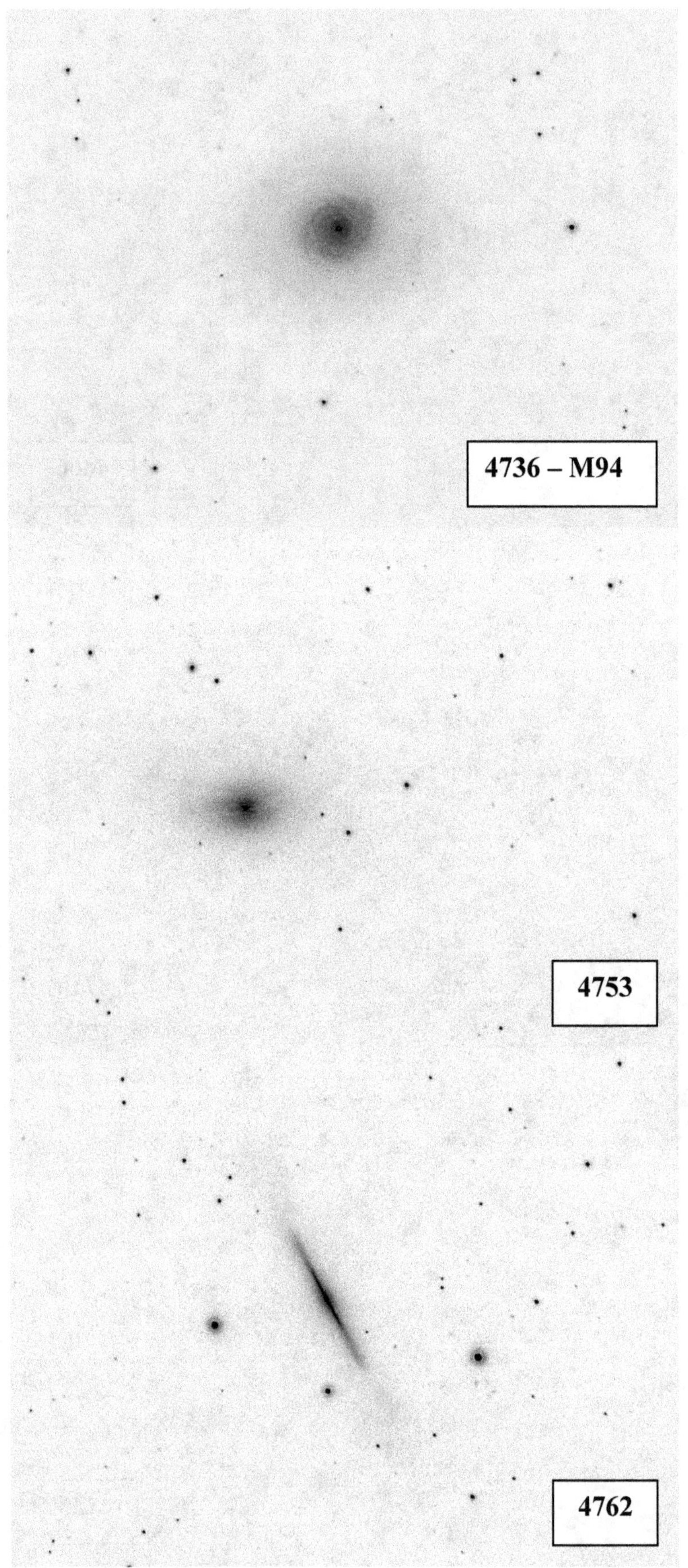
4736 – M94
4753
4762

RA 12h

4826 – M64

4845

4866

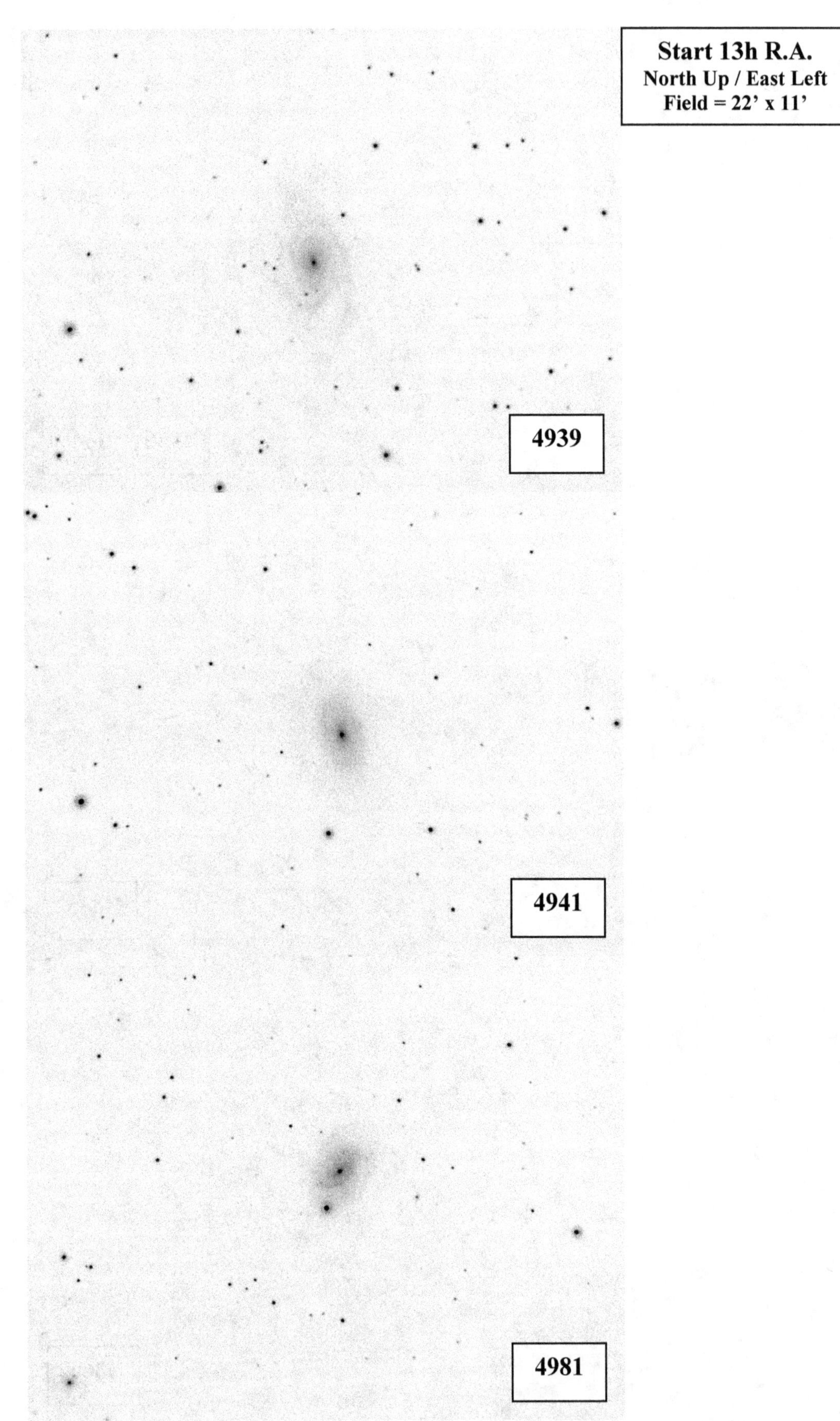
Start 13h R.A.
North Up / East Left
Field = 22' x 11'
4939
4941
4981

RA 13h

4995

5005

5033

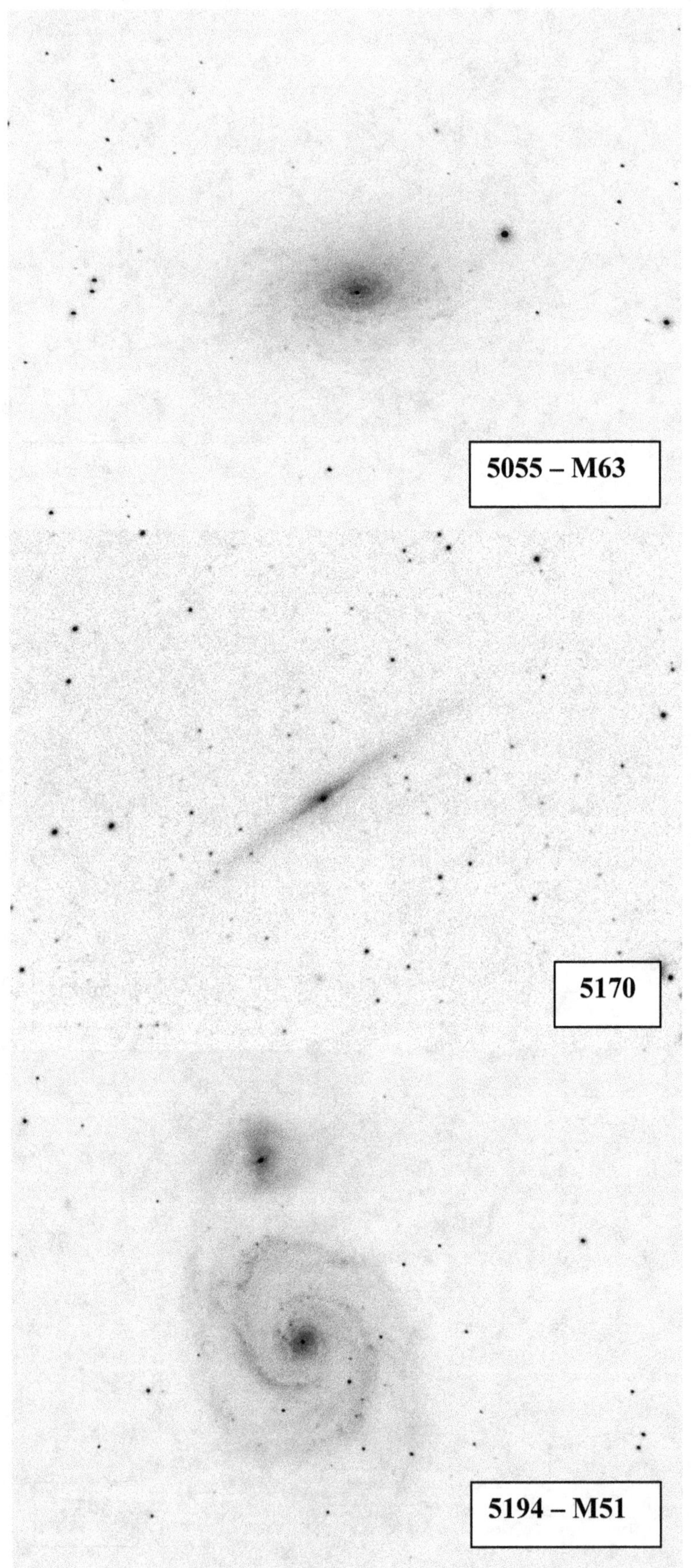
5055 – M63
5170
5194 – M51

RA 13h

5218-5216

5236 – M83

5247

RA 13h

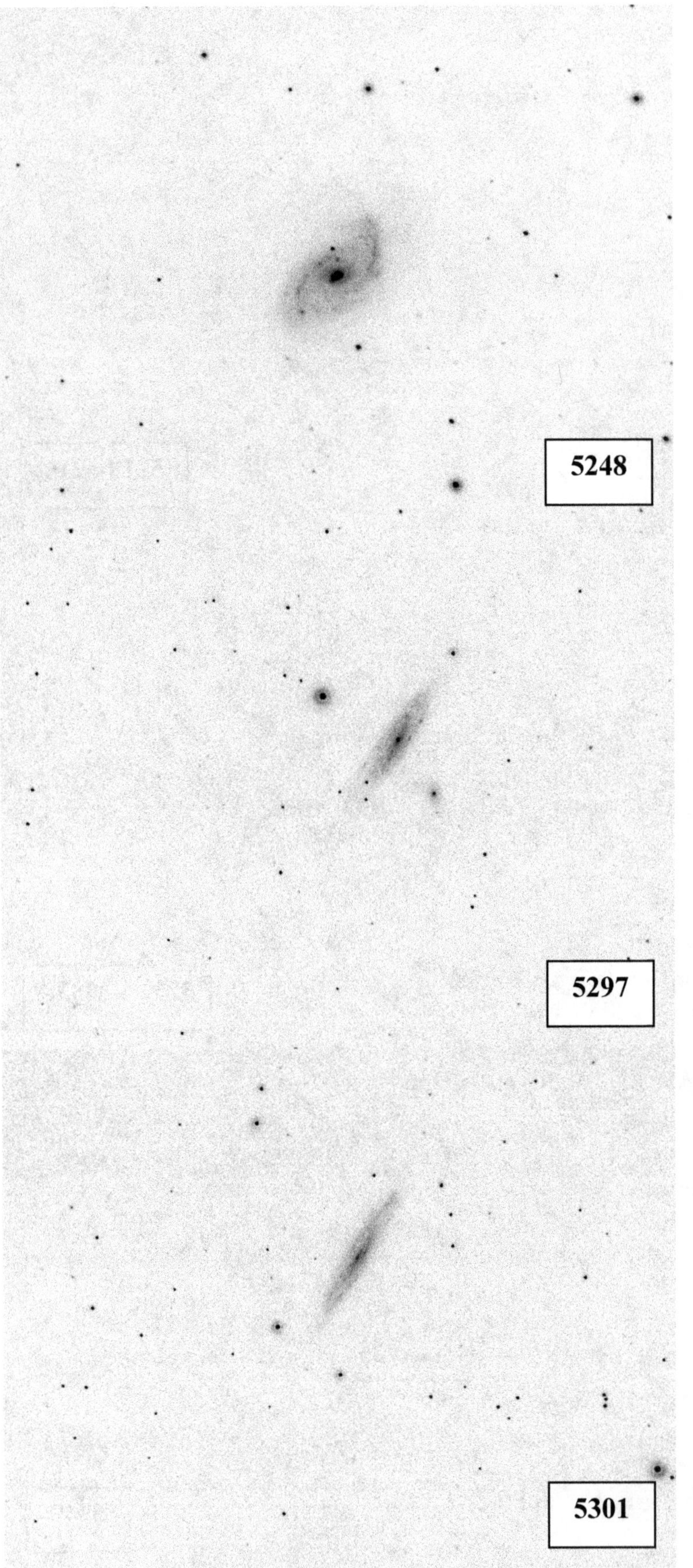

RA 13h

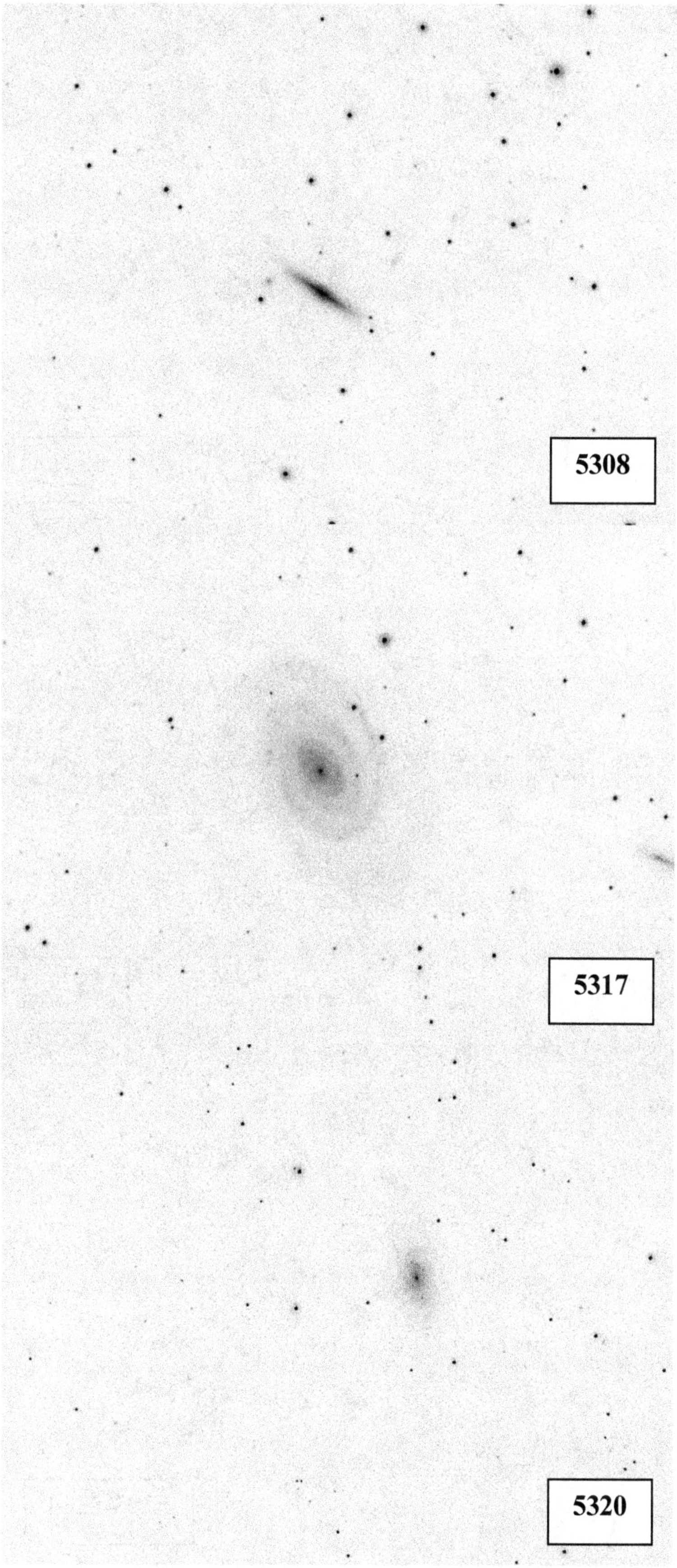

5308

5317

5320

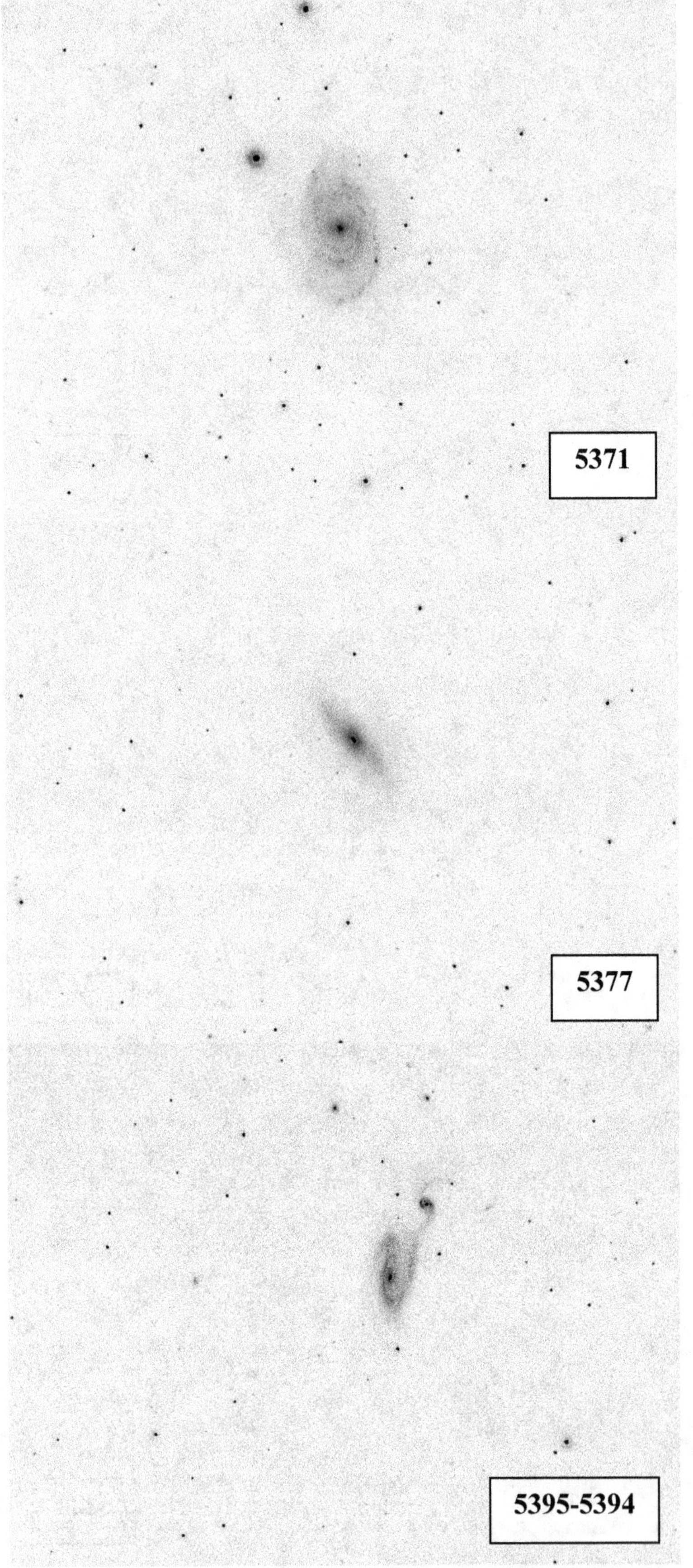
5371
5377
5395-5394

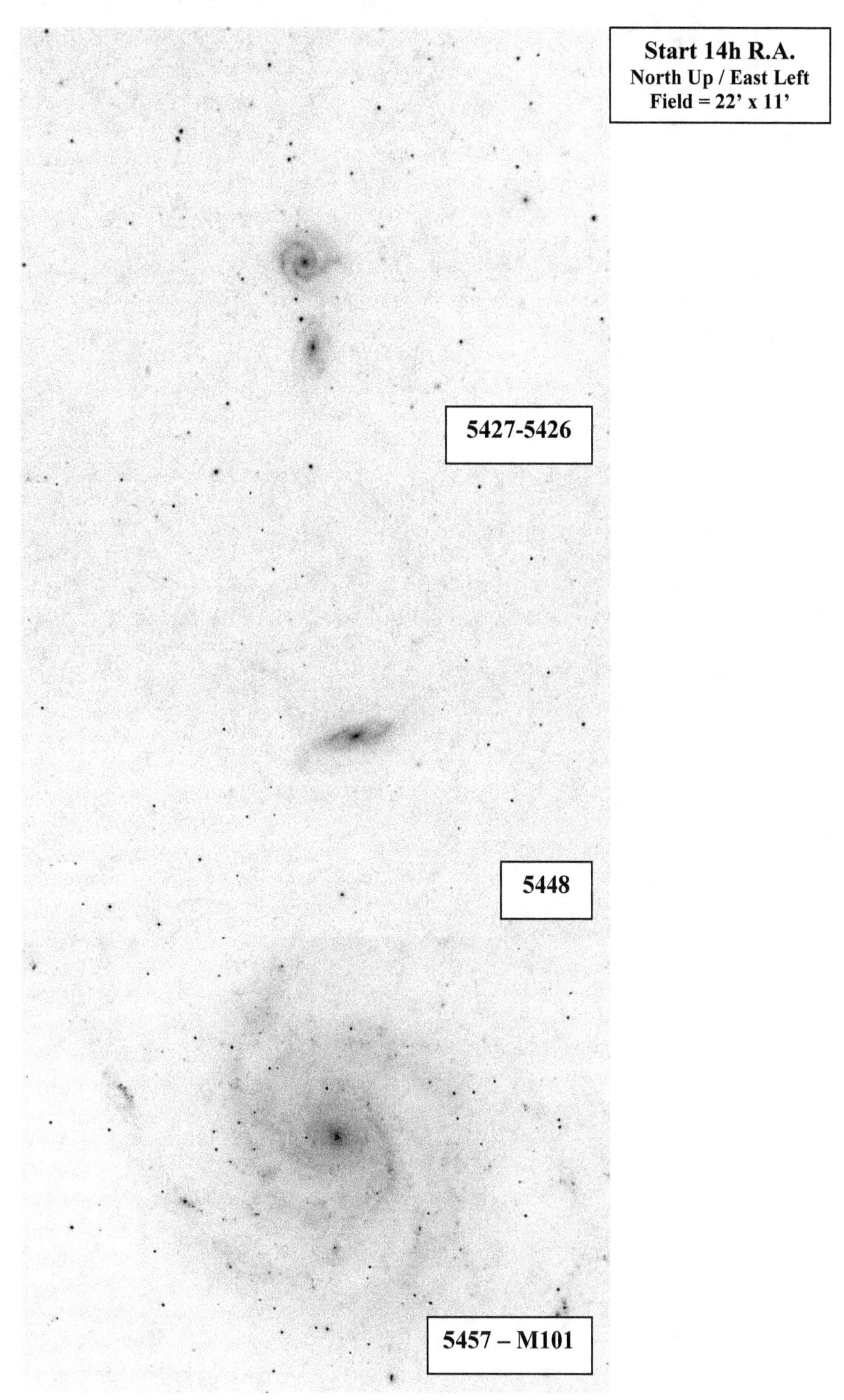
Start 14h R.A.
North Up / East Left
Field = 22' x 11'
5427-5426
5448
5457 – M101

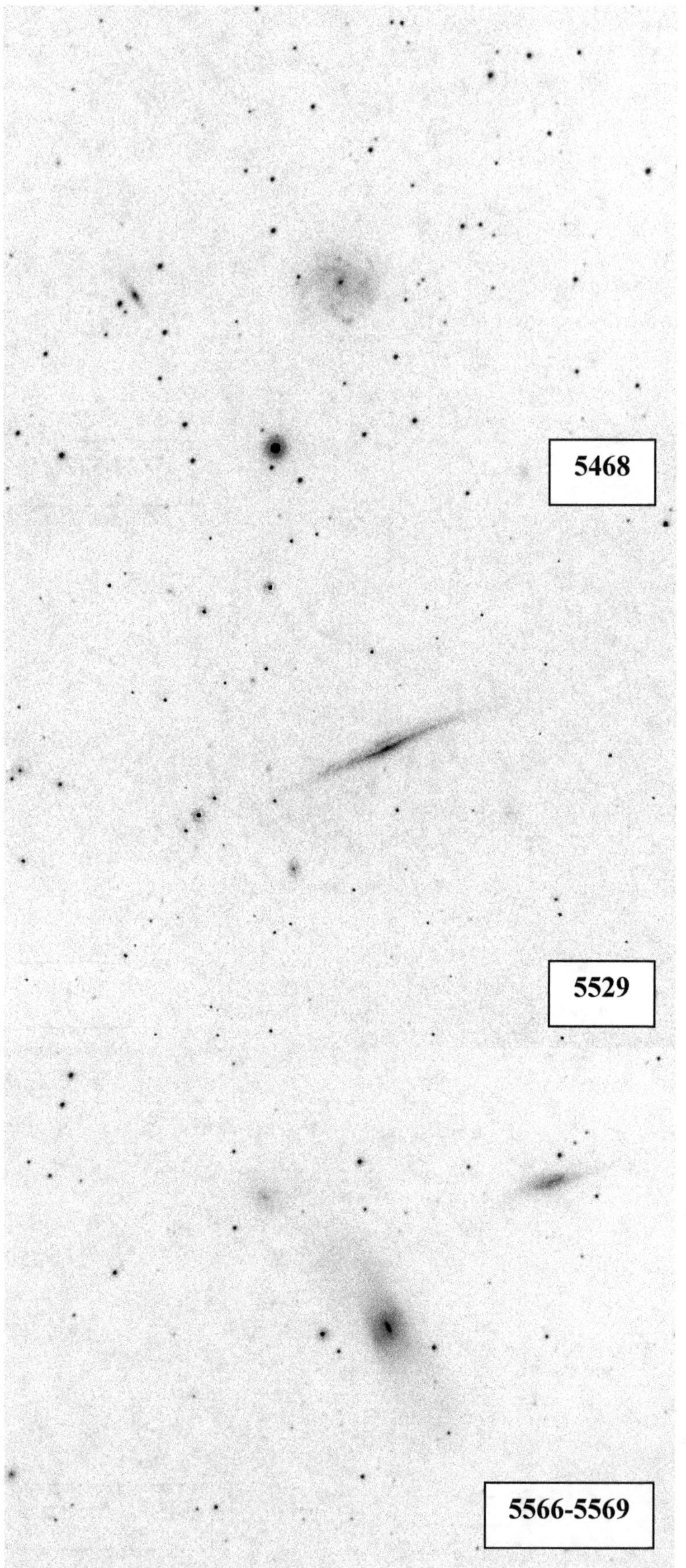
5468
5529
5566-5569

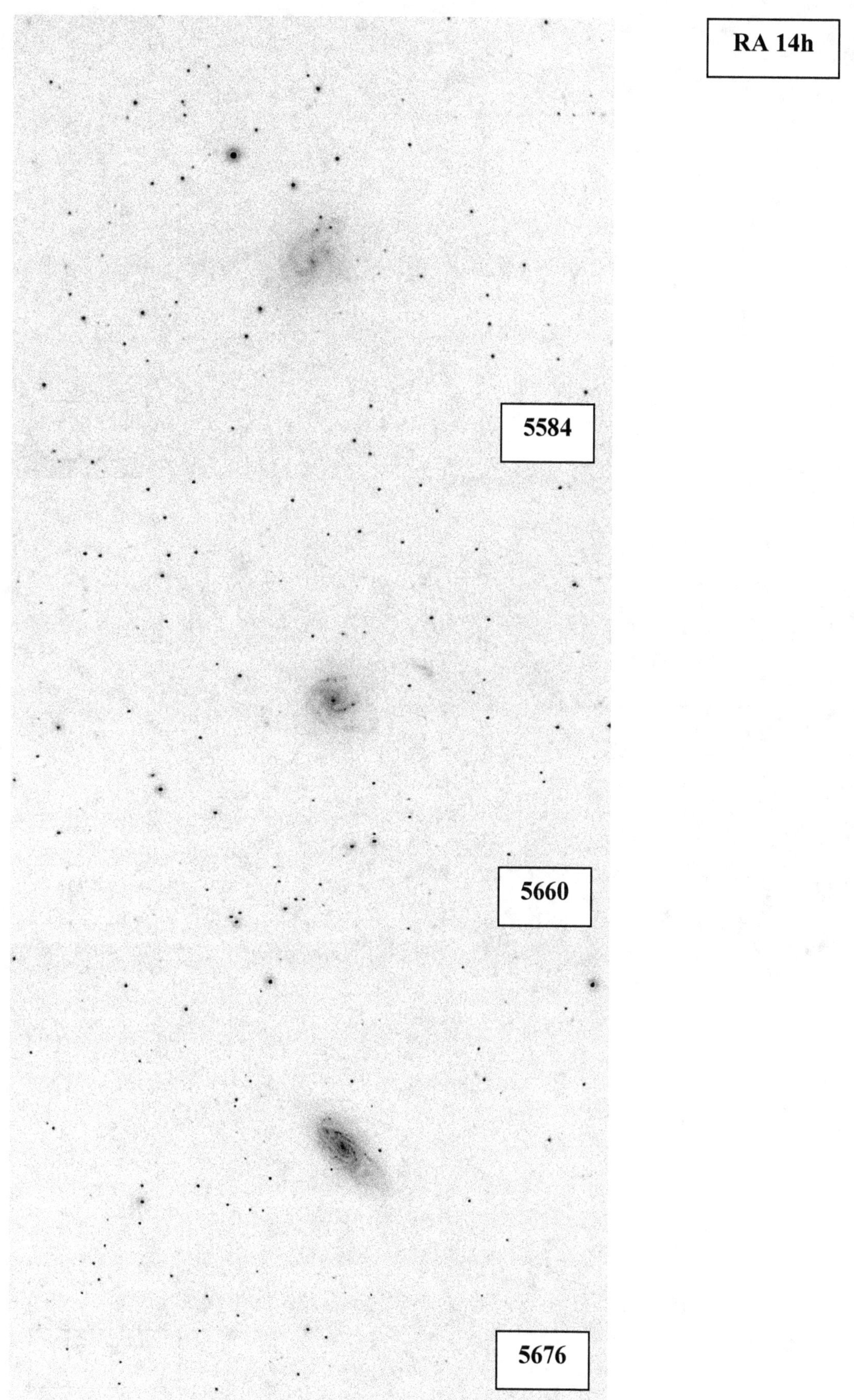
RA 14h
5584
5660
5676

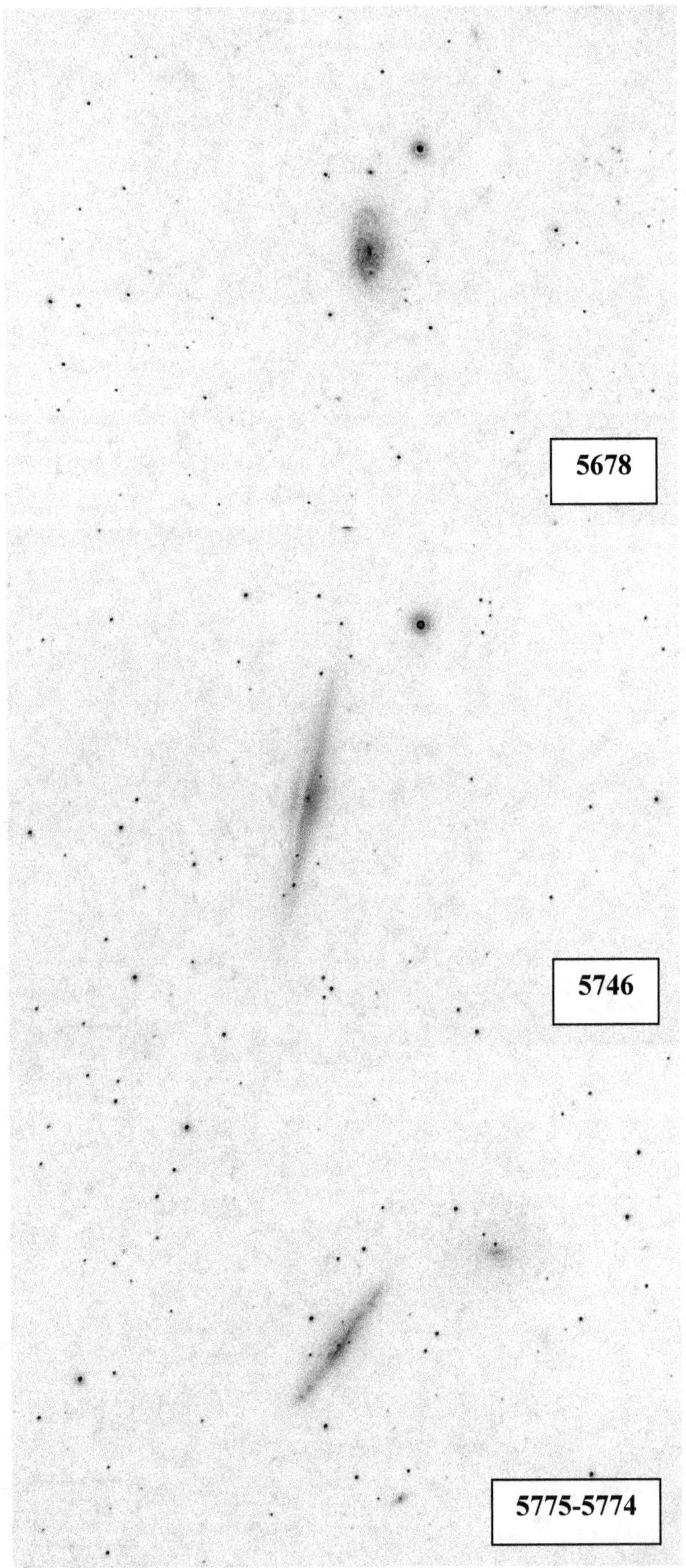
5678
5746
5775-5774

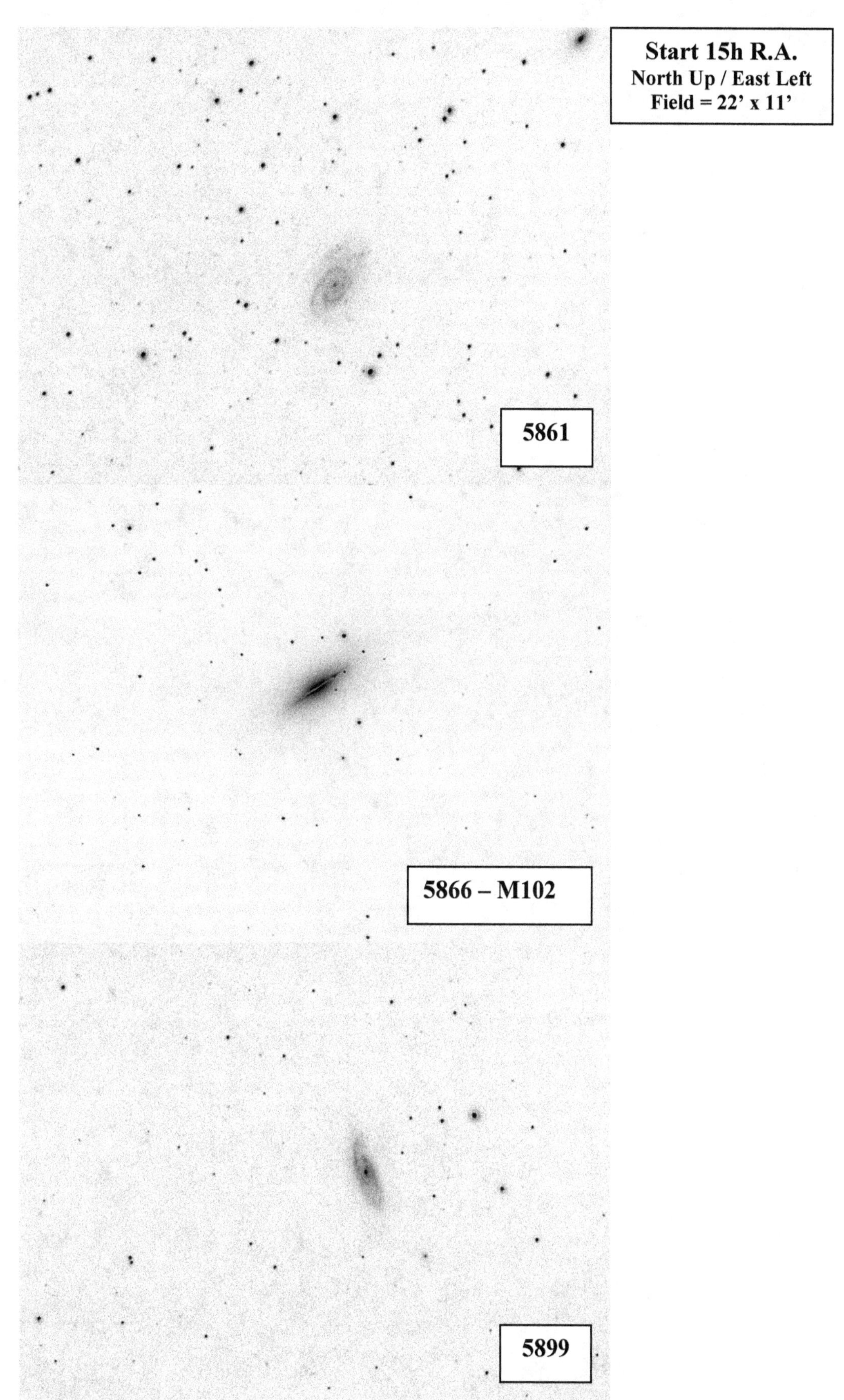
Start 15h R.A.
North Up / East Left
Field = 22' x 11'
5861
5866 – M102
5899

RA 15h

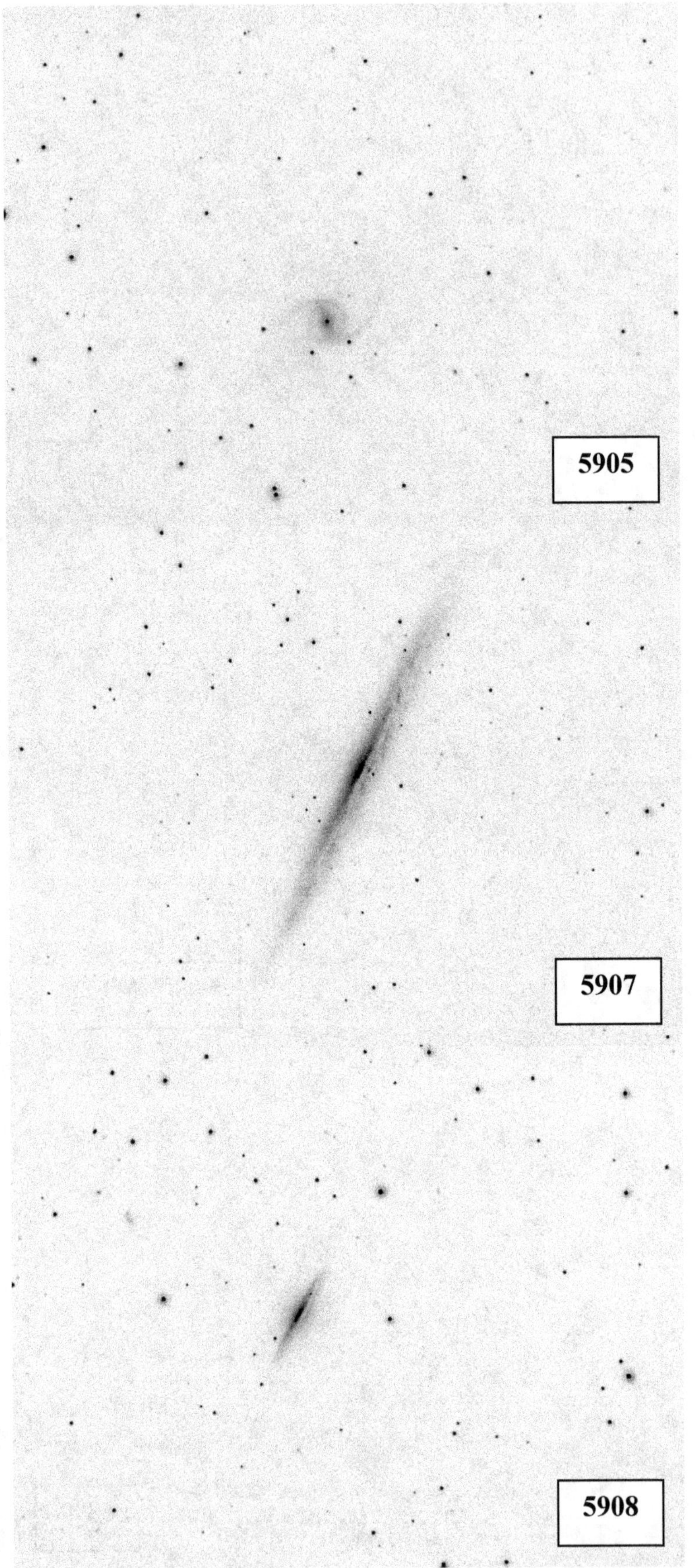

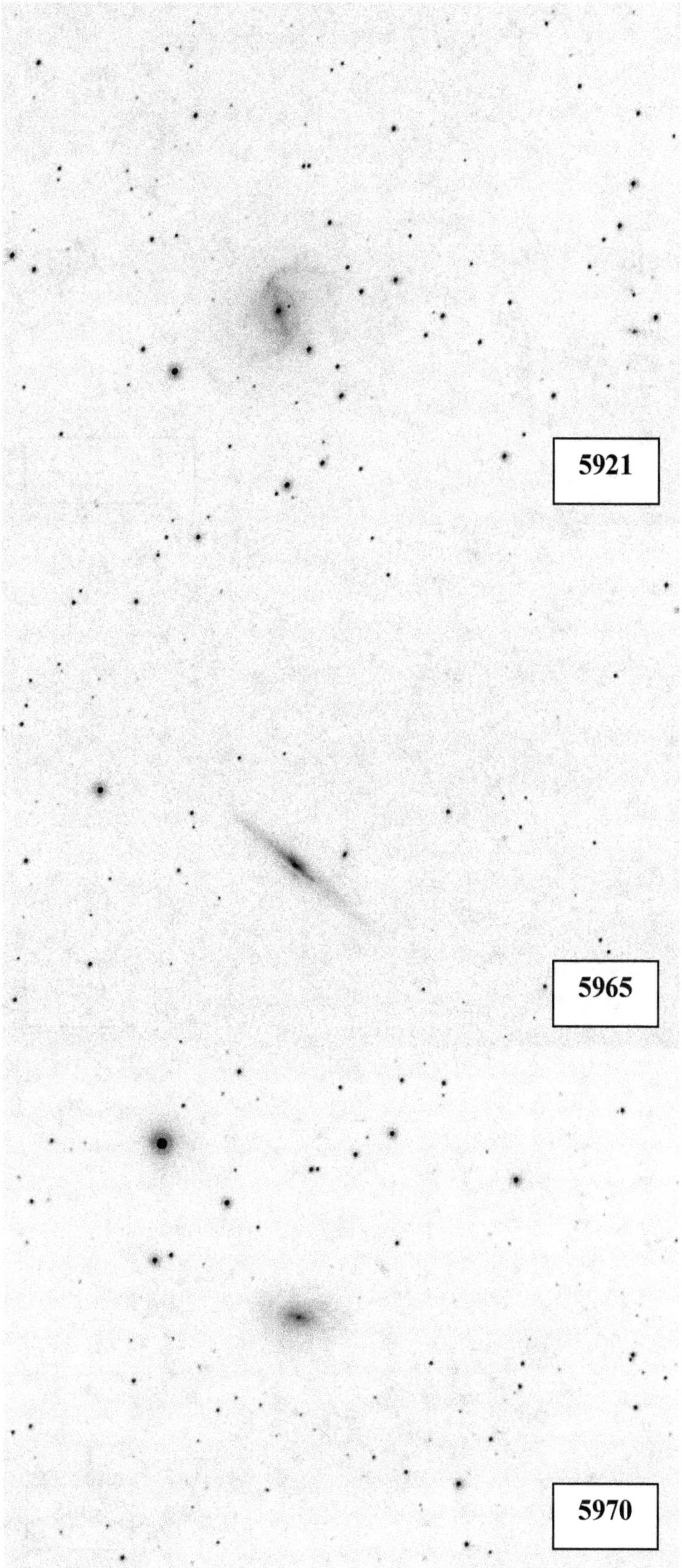

RA 15h

5921

5965

5970

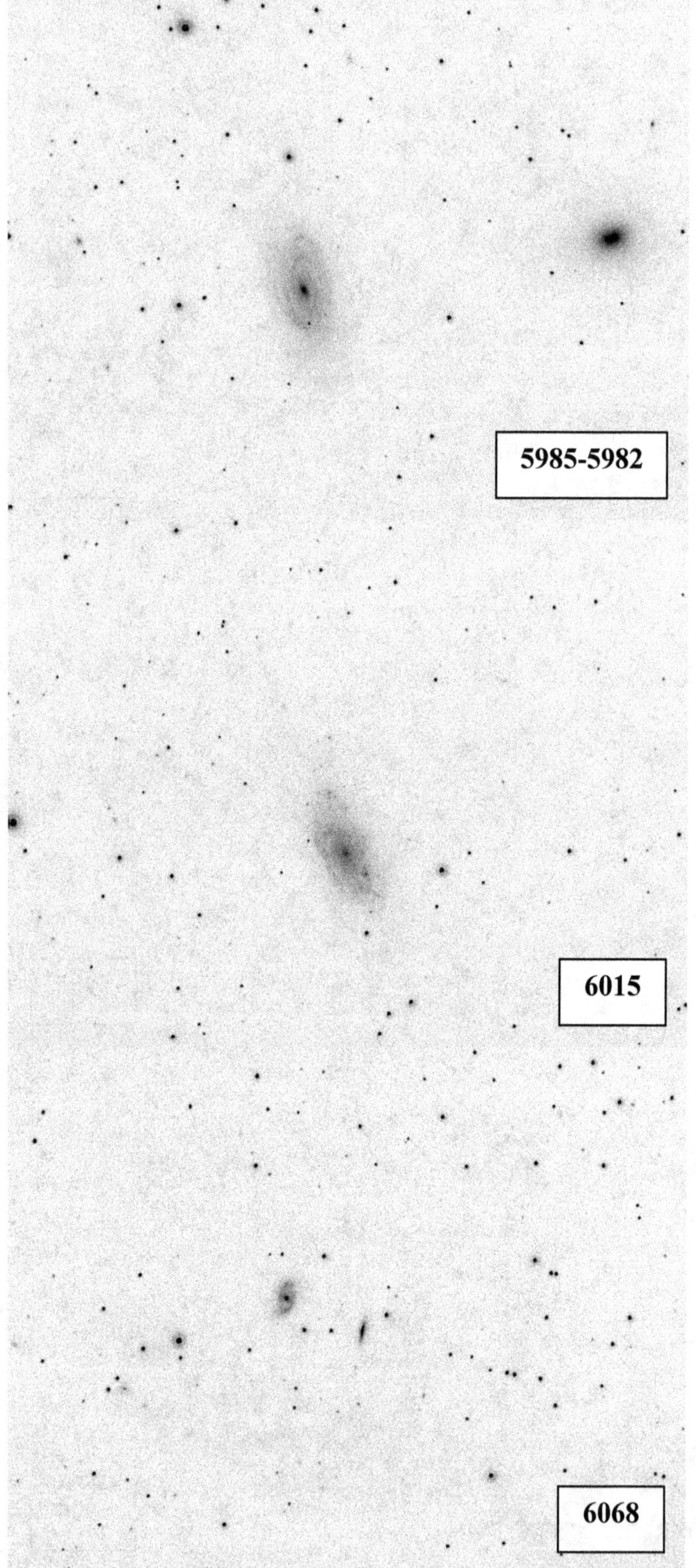

RA 15h

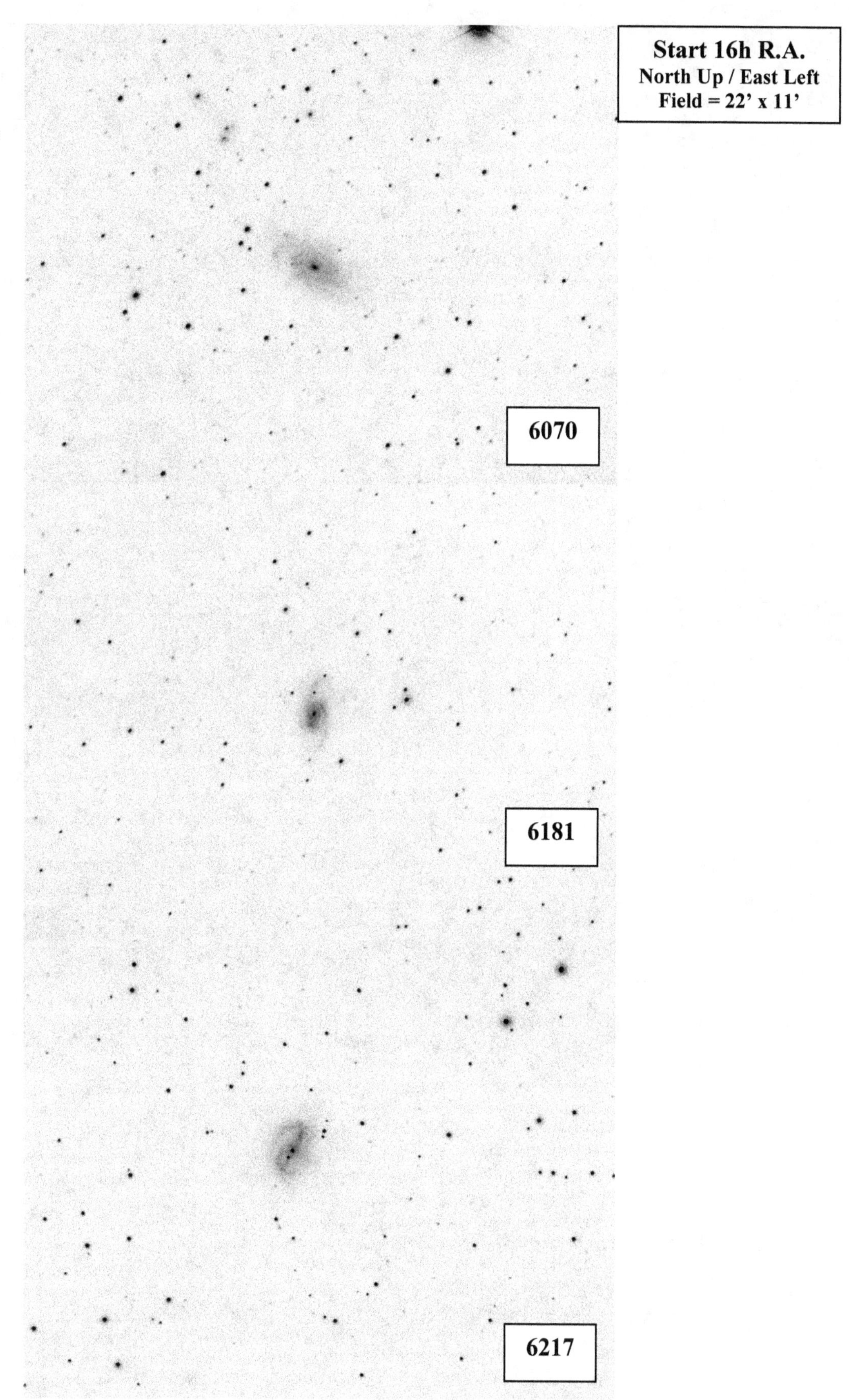
Start 16h R.A.
North Up / East Left
Field = 22' x 11'
6070
6181
6217

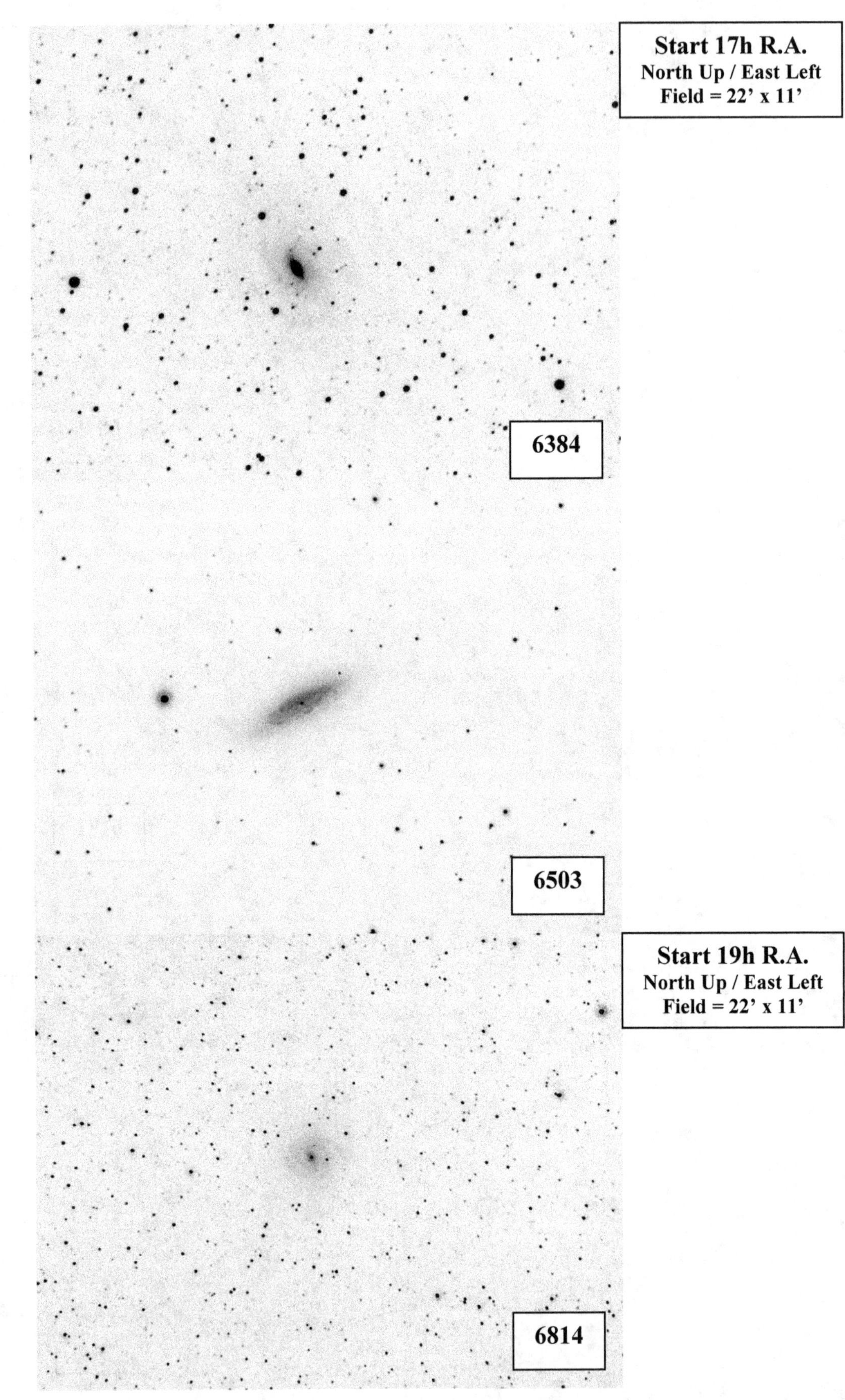
Start 17h R.A.
North Up / East Left
Field = 22' x 11'
6384
6503
Start 19h R.A.
North Up / East Left
Field = 22' x 11'
6814

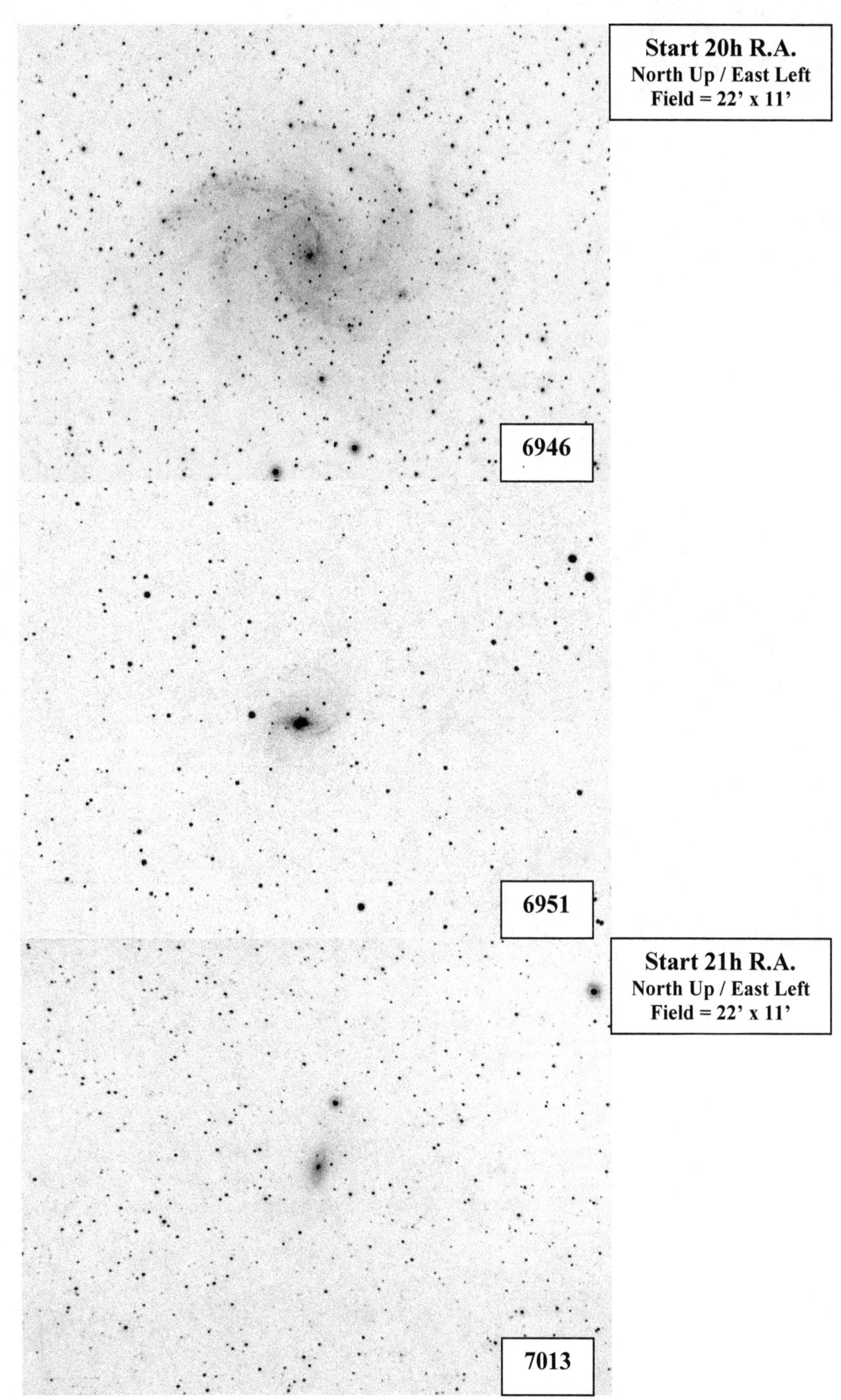
Start 20h R.A.
North Up / East Left
Field = 22' x 11'
6946
6951
Start 21h R.A.
North Up / East Left
Field = 22' x 11'
7013

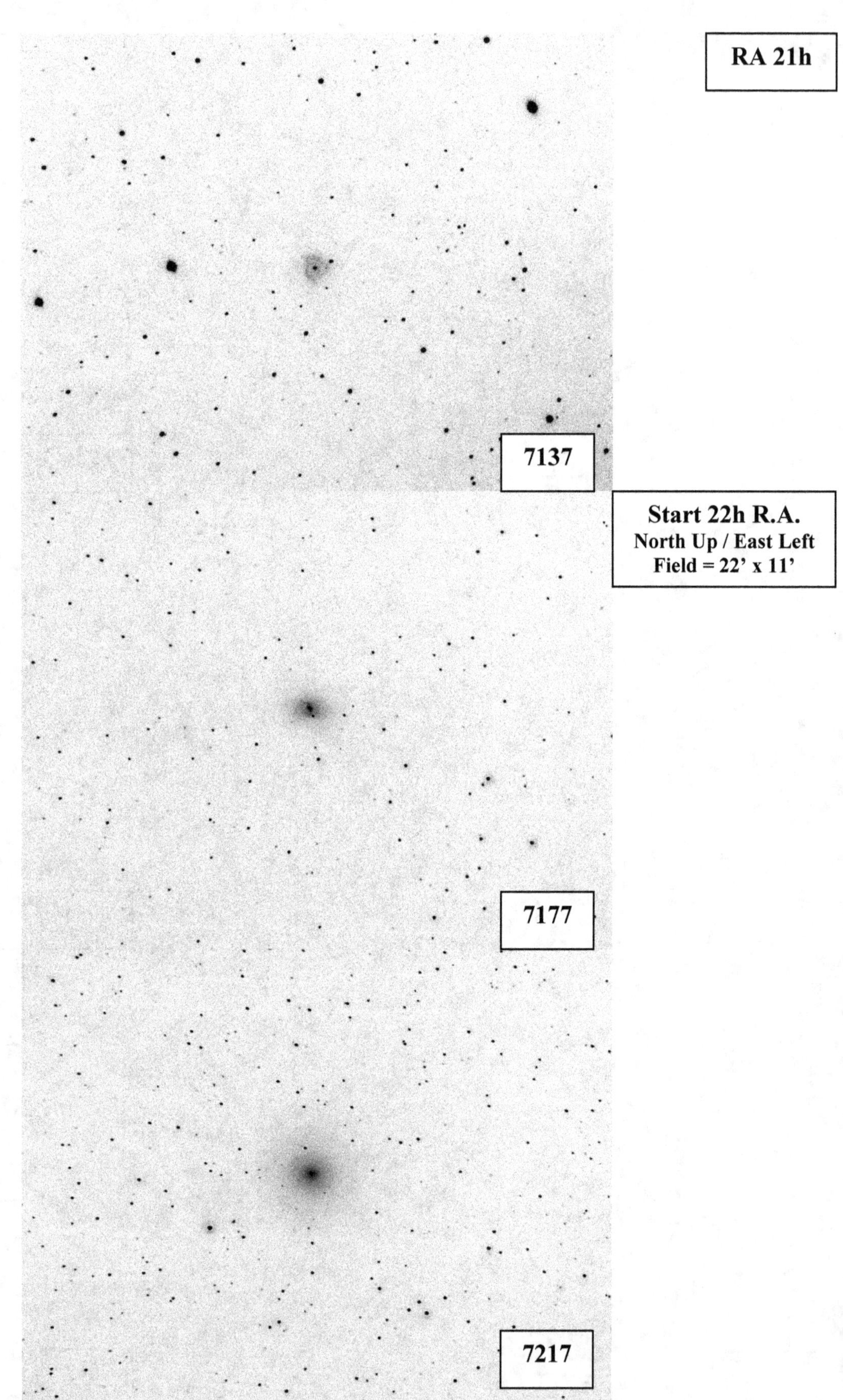
RA 21h
7137
Start 22h R.A.
North Up / East Left
Field = 22' x 11'
7177
7217

RA 22h

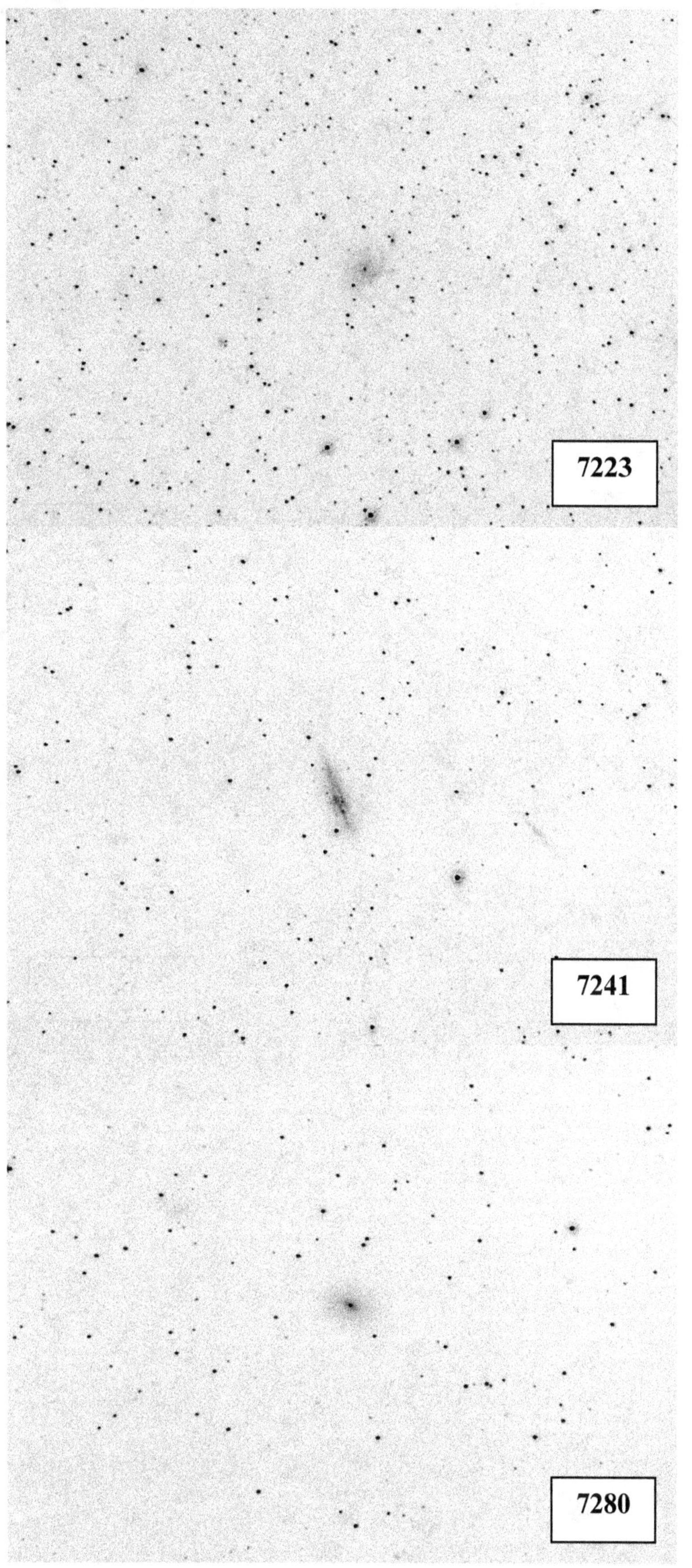

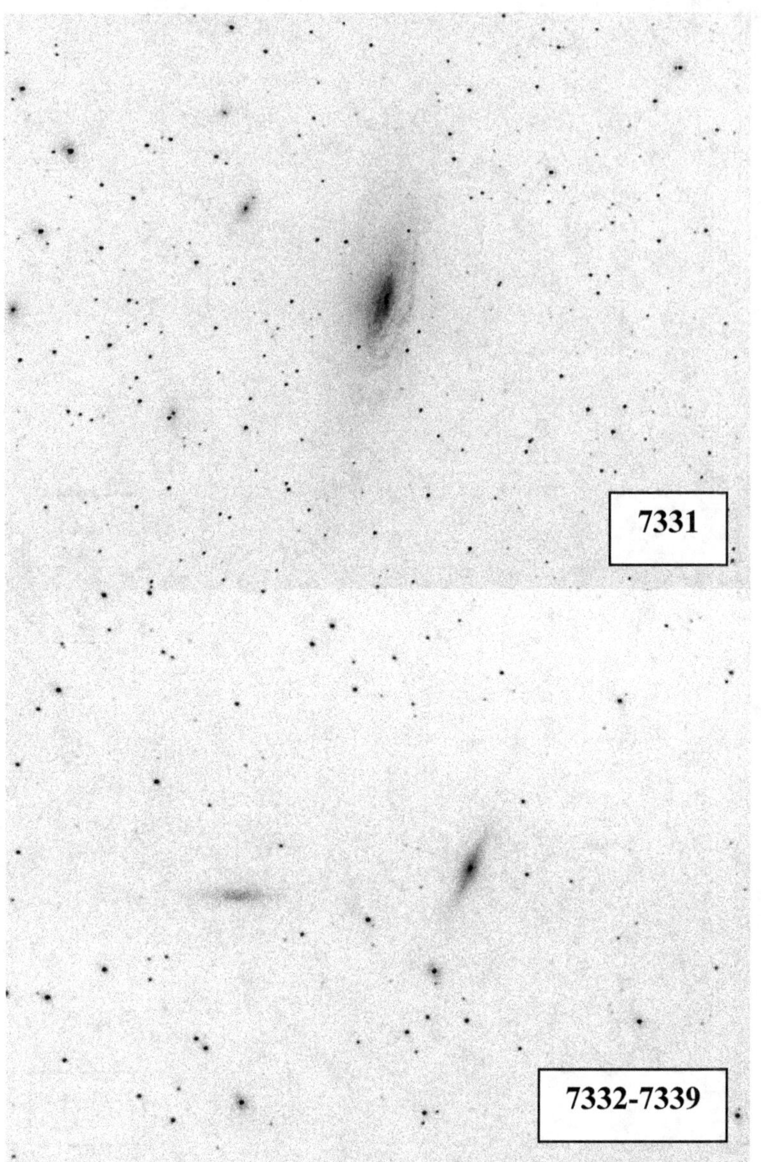
7331
7332-7339

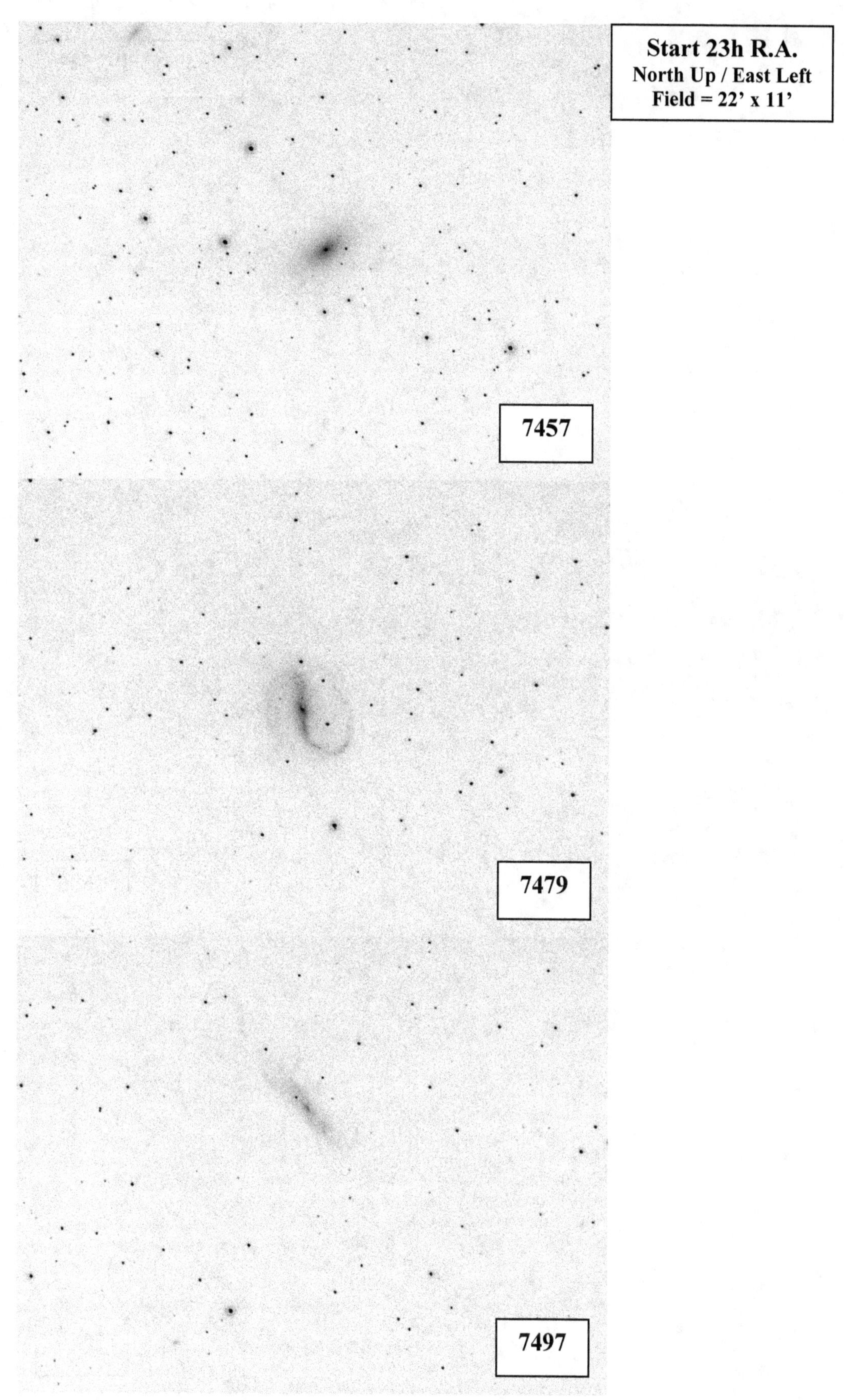
Start 23h R.A.
North Up / East Left
Field = 22' x 11'
7457
7479
7497

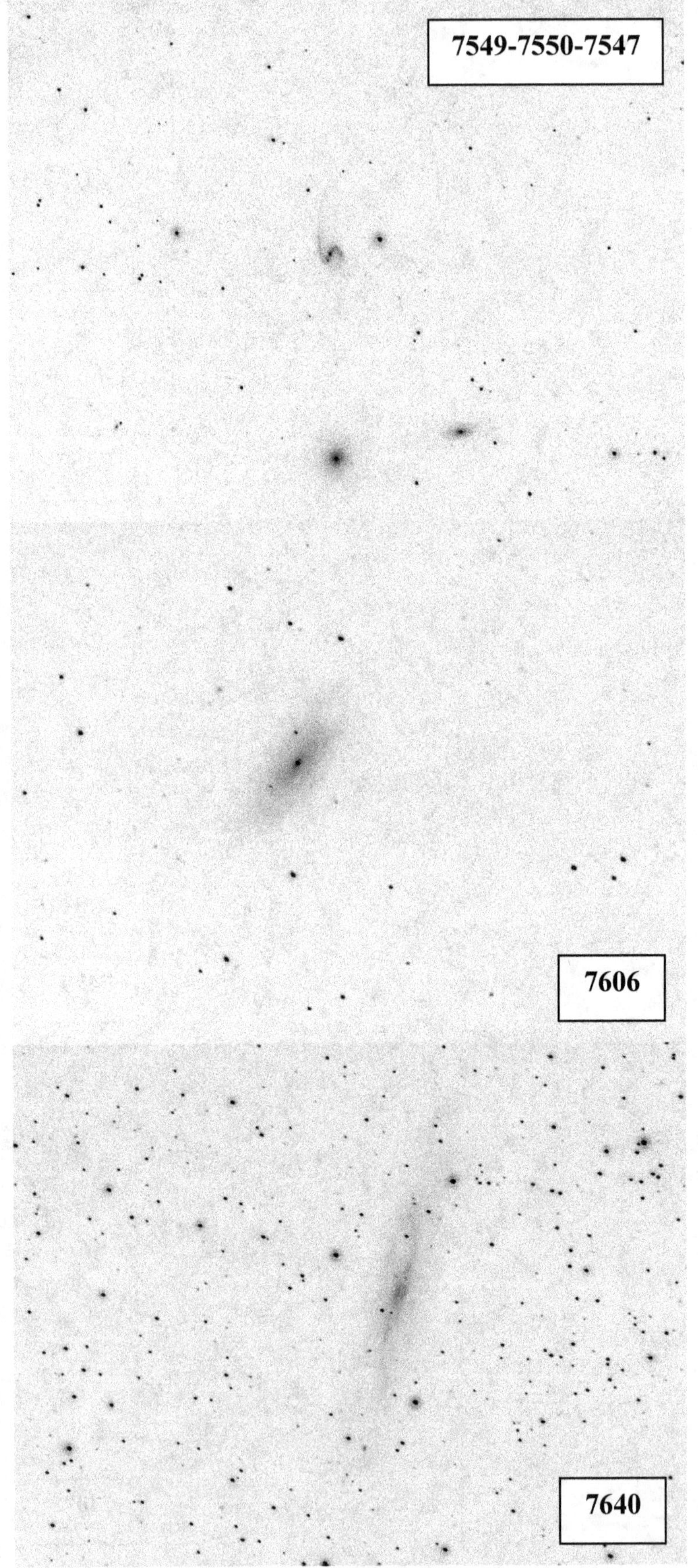
7549-7550-7547
7606
7640

RA 23h

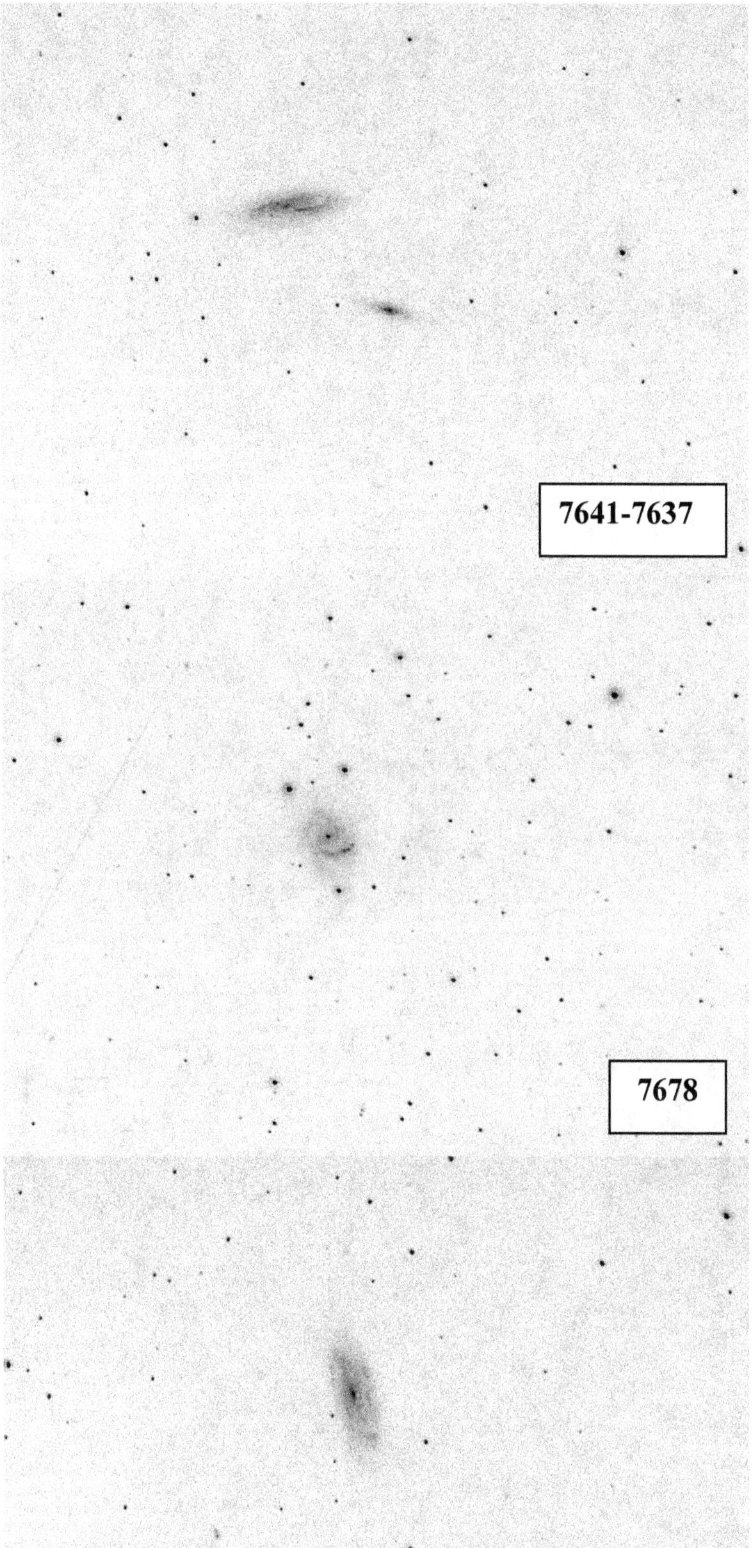

7641-7637

7678

7721

RA 23h

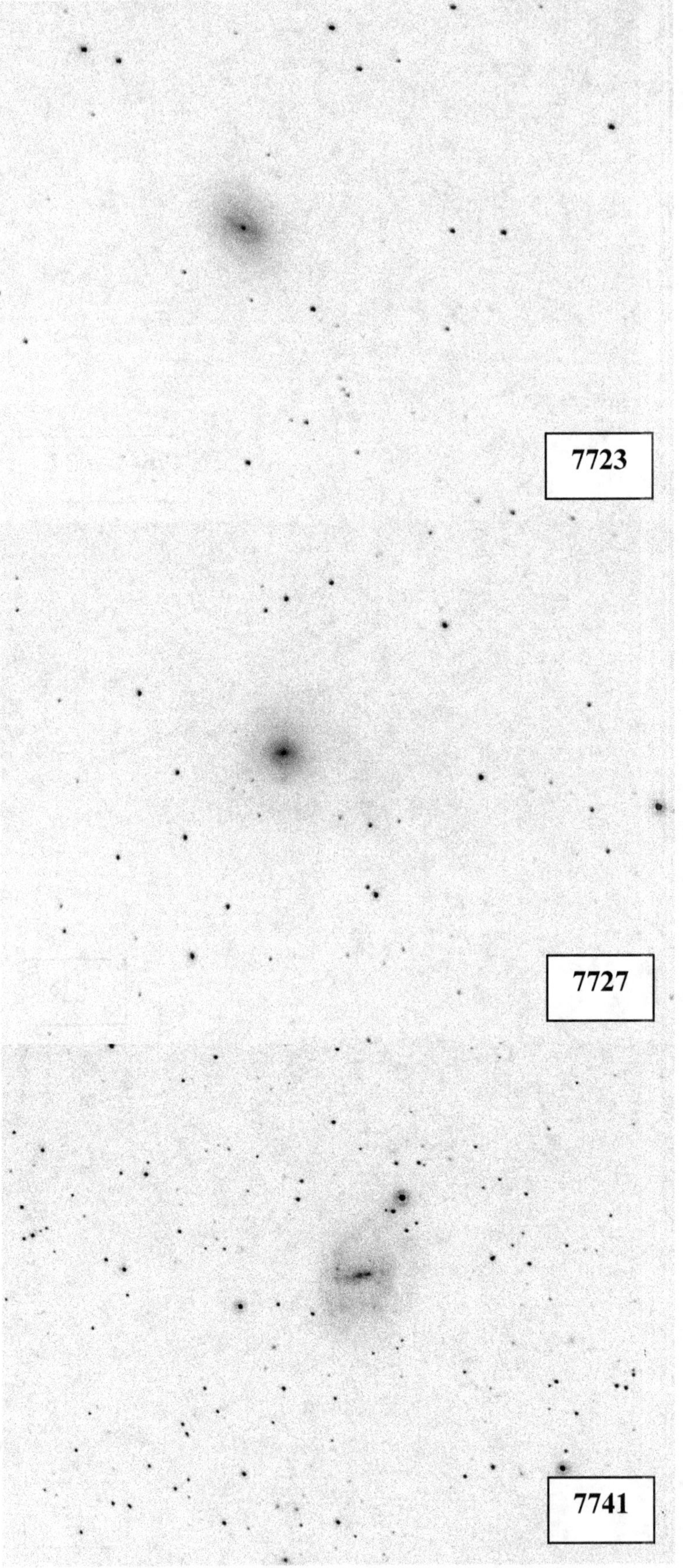

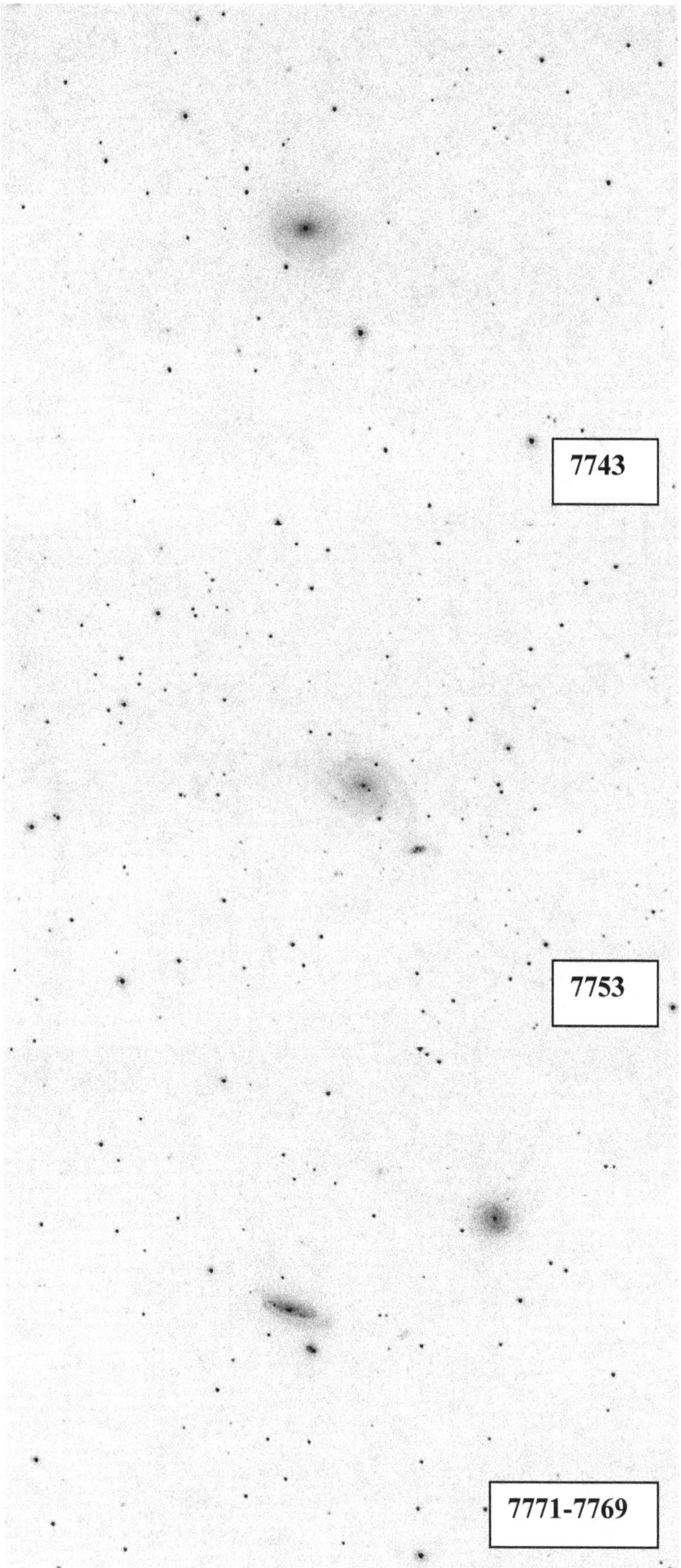

RA 23h

7743

7753

7771-7769

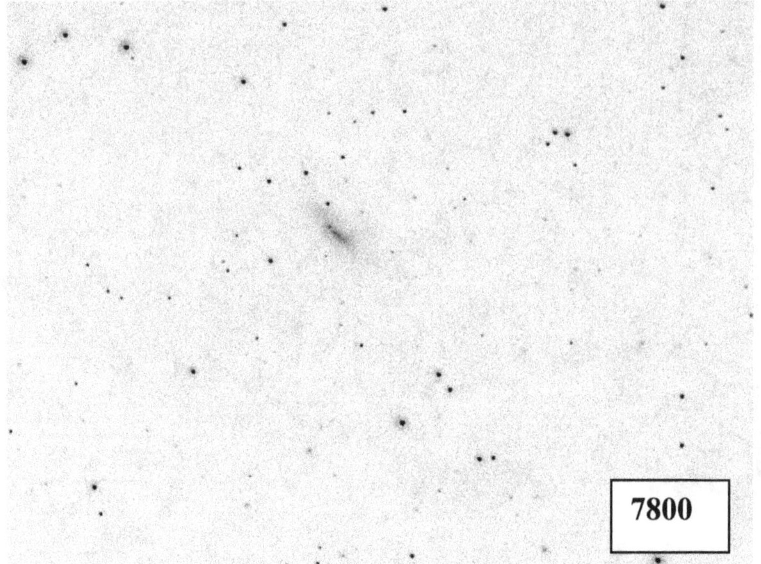
7800

APPENDIX I – MESSIER 31 – The Andromeda Galaxy

Messier 31 (NGC 224) is such an unusual and seemingly easy study target to search for supernovae activity. It is the closest spiral galaxy to our own Milky Way and is even visible to the naked eye on a dark night, high overhead at dark in early winter skies.

However, as we peer at this enticing object telescopically, we realize two factors that make it suddenly known as not-so-easy in terms of study for changes:

1) the galaxy is so close that it cannot be adequately examined for 15th-19th magnitude stars if viewing the entire span of the galaxy at one time – it simply is too large to fit in the CCD field of view of suitably large telescopes for supernova searches and study. Wide field telescopes, capable of capturing the huge span of over three degrees simply cannot reach the faint magnitudes required to detect supernovae, even at this "close distance' of only 2.3 million light years.
2) there are many foreground stars between our view here on earth and the galaxy, these all being the curtains of stars from our own Milky Way through which we peer.

Thus, the CCD images for this galaxy should be broken into segments for study, consistent with all other image scales and fields of view used for all other NGC galaxies for which we search the explosive events.

In the case of the Arkansas Sky Observatories' system, with a field of view of roughly 22 x 15 arc minutes, the total galaxy is broken into THREE study areas: **core**, **northeastern arm**, and **southwestern arm**. This allows proper blinking of each of these three areas and provides adequate image scale for improved detection possibilities.

Messier 31 (ngc224) as taken with the ASO 0.13m TMB f/6 astrographic refractor

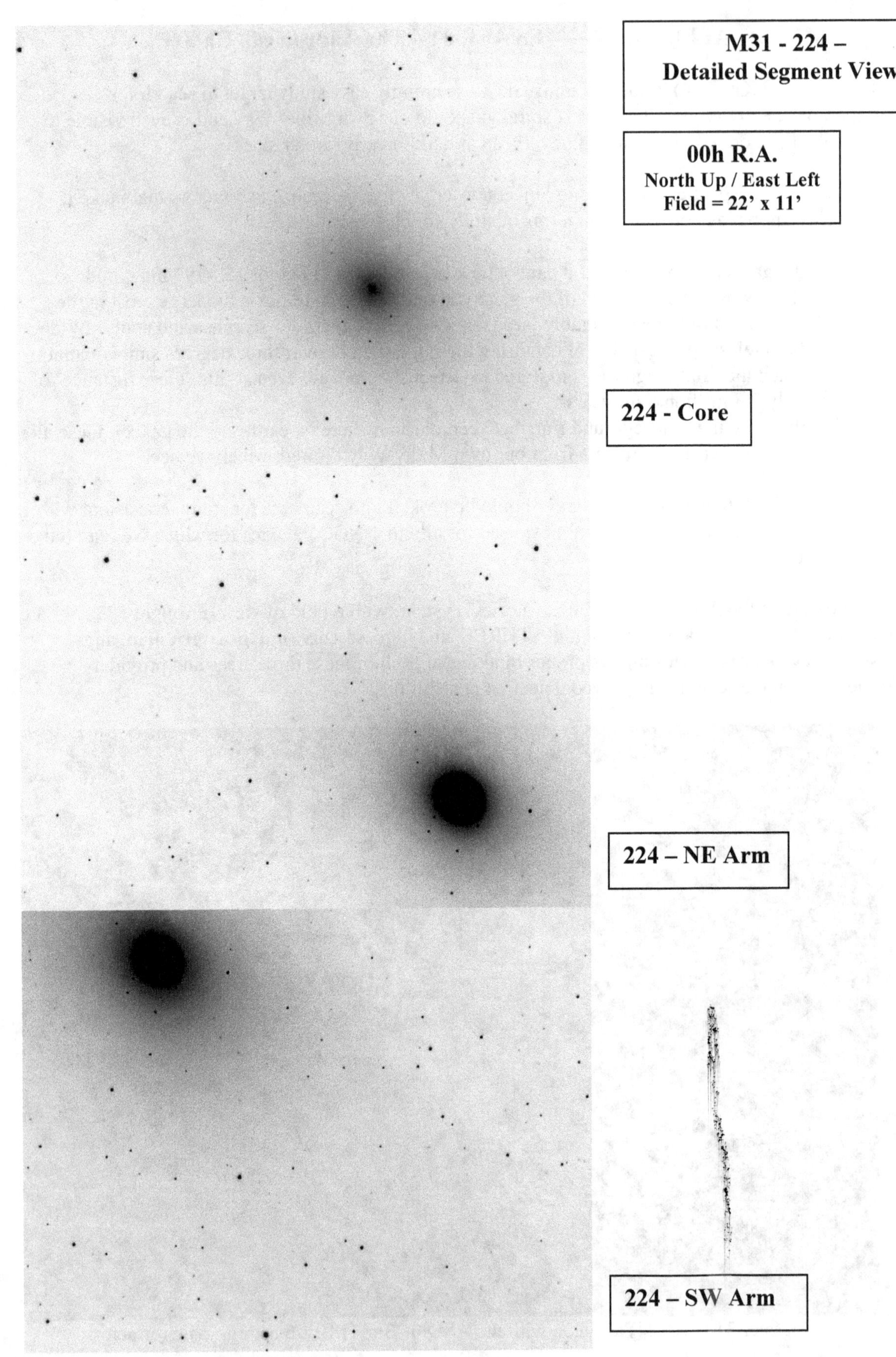
M31 - 224 –
Detailed Segment Views
00h R.A.
North Up / East Left
Field = 22' x 11'
224 - Core
224 – NE Arm
224 – SW Arm

APPENDIX II – Star Color Index, Differential Photometry

The following table provides a guide to the color indexes for main sequence stars of spectral types B0 to M4. The data are taken from Fitzgerald (1970 - A&A 4, 234), who provides UBV data for spectral classes O5 to M8, and from Ducati et al. (2001 - ApJ, 558, 309), who cover B0 to M4. This data is presented for those who wish to do follow-up studies on supernovae and novae discoveries using the UBVRI filtered photometric differential spectroscopy method to assess changes in star color as the nova/supernova event progresses and evolves from discovery through decline.

Sp. Type	U-B	(B-V)	$(V-R)_C$	$(V-I)_C$	(V-J)	(V-H)	(V-K)	(V-L)	(V-M)	(V-N)	$(V-R)_J$	$(V-I)_J$
B0.0	-1.08	-0.30	-0.19	-0.31	-0.80	-0.92	-0.97	-1.13	-1.00	-9.99	-0.22	-0.44
B0.5	-1.00	-0.28	-0.18	-0.31	-0.77	-0.89	-0.95	-1.11	-0.99	-9.99	-0.20	-0.43
B1.0	-0.95	-0.26	-0.16	-0.30	-0.73	-0.85	-0.93	-1.08	-0.96	-9.99	-0.18	-0.42
B1.5	-0.88	-0.25	-0.15	-0.29	-0.70	-0.82	-0.91	-1.05	-0.94	-9.99	-0.17	-0.41
B2.0	-0.81	-0.24	-0.14	-0.29	-0.67	-0.79	-0.89	-1.02	-0.92	-9.99	-0.15	-0.40
B2.5	-0.72	-0.22	-0.13	-0.28	-0.64	-0.76	-0.86	-0.97	-0.88	-0.96	-0.14	-0.39
B3.0	-0.68	-0.20	-0.12	-0.27	-0.60	-0.72	-0.82	-0.92	-0.84	-0.91	-0.13	-0.38
B3.5	-0.65	-0.19	-0.12	-0.26	-0.58	-0.70	-0.80	-0.90	-0.82	-0.87	-0.12	-0.37
B4.0	-0.63	-0.18	-0.11	-0.25	-0.56	-0.68	-0.77	-0.86	-0.79	-0.84	-0.11	-0.35
B4.5	-0.61	-0.17	-0.11	-0.24	-0.54	-0.65	-0.74	-0.83	-0.76	-0.80	-0.11	-0.34
B5.0	-0.58	-0.16	-0.10	-0.24	-0.51	-0.62	-0.71	-0.78	-0.73	-0.75	-0.10	-0.33
B6.0	-0.49	-0.14	-0.10	-0.21	-0.46	-0.57	-0.64	-0.70	-0.65	-0.66	-0.09	-0.29
B7.0	-0.43	-0.13	-0.09	-0.19	-0.41	-0.51	-0.57	-0.61	-0.58	-0.58	-0.08	-0.26
B7.5	-0.40	-0.12	-0.09	-0.17	-0.39	-0.48	-0.54	-0.57	-0.54	-0.53	-0.08	-0.24
B8.0	-0.36	-0.11	-0.08	-0.16	-0.36	-0.45	-0.49	-0.52	-0.49	-0.48	-0.07	-0.22
B8.5	-0.27	-0.09	-0.08	-0.13	-0.31	-0.40	-0.43	-0.43	-0.42	-0.39	-0.07	-0.18
B9.0	-0.18	-0.07	-0.07	-0.10	-0.26	-0.34	-0.33	-0.34	-0.34	-0.30	-0.06	-0.14
B9.5	-0.10	-0.04	-0.05	-0.08	-0.22	-0.29	-0.26	-0.27	-0.26	-0.22	-0.03	-0.11
A0.0	-0.02	-0.01	-0.04	-0.04	-0.16	-0.19	-0.17	-0.18	-0.18	-0.14	-0.01	-0.05
A1.0	0.01	0.02	-0.02	-0.02	-0.11	-0.12	-0.11	-0.12	-0.13	-0.08	0.01	-0.03
A2.0	0.05	0.05	-0.01	0.00	-0.07	-0.04	-0.05	-0.07	-0.08	-0.02	0.03	0.00
A3.0	0.08	0.08	0.01	0.02	-0.02	0.03	0.01	-0.01	-0.02	0.03	0.05	0.02
A4.0	0.09	0.12	0.02	0.05	0.03	0.11	0.08	0.05	-0.04	0.09	0.07	0.07
A5.0	0.09	0.15	0.04	0.09	0.09	0.19	0.15	0.12	0.10	0.16	0.10	0.12
A6.0	0.10	0.17	0.05	0.12	0.13	0.30	0.21	0.17	0.15	0.21	0.11	0.16
A7.0	0.10	0.20	0.06	0.15	0.18	0.32	0.27	0.23	0.20	0.26	0.13	0.19
A8.0	0.09	0.27	0.09	0.20	0.25	0.42	0.36	0.33	0.29	0.34	0.16	0.26
A9.0	0.08	0.30	0.10	0.24	0.31	0.49	0.44	0.41	0.36	0.41	0.18	0.31
F0.0	0.03	0.32	0.12	0.28	0.37	0.57	0.52	0.49	0.43	0.48	0.21	0.36
F1.0	0.00	0.34	0.14	0.31	0.43	0.64	0.58	0.57	0.49	0.54	0.23	0.40
F2.0	0.00	0.35	0.15	0.35	0.48	0.71	0.66	0.66	0.56	0.60	0.25	0.45

F5.0	-0.02	0.45	0.21	0.44	0.67	0.93	0.89	0.90	0.77	0.80	0.33	0.57
F8.0	0.02	0.53	0.24	0.50	0.79	1.06	1.03	1.06	0.91	0.91	0.37	0.64
G0.0	0.06	0.60	0.27	0.54	0.87	1.15	1.14	1.18	1.01	1.01	0.41	0.70
G2.0	0.09	0.63	0.30	0.58	0.97	1.25	1.26	1.31	1.12	1.11	0.45	0.75
G3.0	0.12	0.65	0.30	0.59	0.98	1.27	1.28	1.33	1.14	1.13	0.45	0.76
G5.0	0.20	0.68	0.31	0.61	1.02	1.31	1.32	1.38	1.18	1.17	0.47	0.78
G8.0	0.30	0.74	0.35	0.66	1.14	1.44	1.47	1.55	1.34	1.30	0.52	0.85
K0.0	0.44	0.81	0.42	0.75	1.34	1.67	1.74	1.85	1.61	1.54	0.61	0.97
K1.0	0.48	0.86	0.46	0.82	1.46	1.80	1.89	2.02	1.78	1.68	0.67	1.05
K2.0	0.67	0.92	0.50	0.89	1.60	1.94	2.06	2.21	1.97	1.84	0.73	1.14
K3.0	0.73	0.95	0.55	0.97	1.73	2.09	2.23	2.40	2.17	2.01	0.80	1.25
K4.0	1.00	1.00	0.60	1.04	1.84	2.22	2.38	2.57	2.36	2.15	0.86	1.34
K5.0	1.06	1.15	0.68	1.20	2.04	2.46	2.66	2.87	2.71	2.44	0.97	1.54
K7.0	1.21	1.33	0.62	1.45	2.30	2.78	3.01	3.25	3.21	2.83	1.13	1.86
M0.0	1.23	1.37	0.70	1.67	2.49	3.04	3.29	3.54	3.65	3.16	1.26	2.15
M1.0	1.18	1.47	0.76	1.84	2.61	3.22	3.47	3.72	3.95	3.39	1.36	2.36
M2.0	1.15	1.47	0.83	2.06	2.74	3.42	3.67	3.92	4.31	3.66	1.46	2.62
M3.0	1.17	1.50	0.89	2.24	2.84	3.58	3.83	4.08	4.62	3.89	1.56	2.84
M4.0	1.07	1.52	0.94	2.43	2.93	3.74	3.98	4.22	4.93	4.11	1.65	3.07

* * *

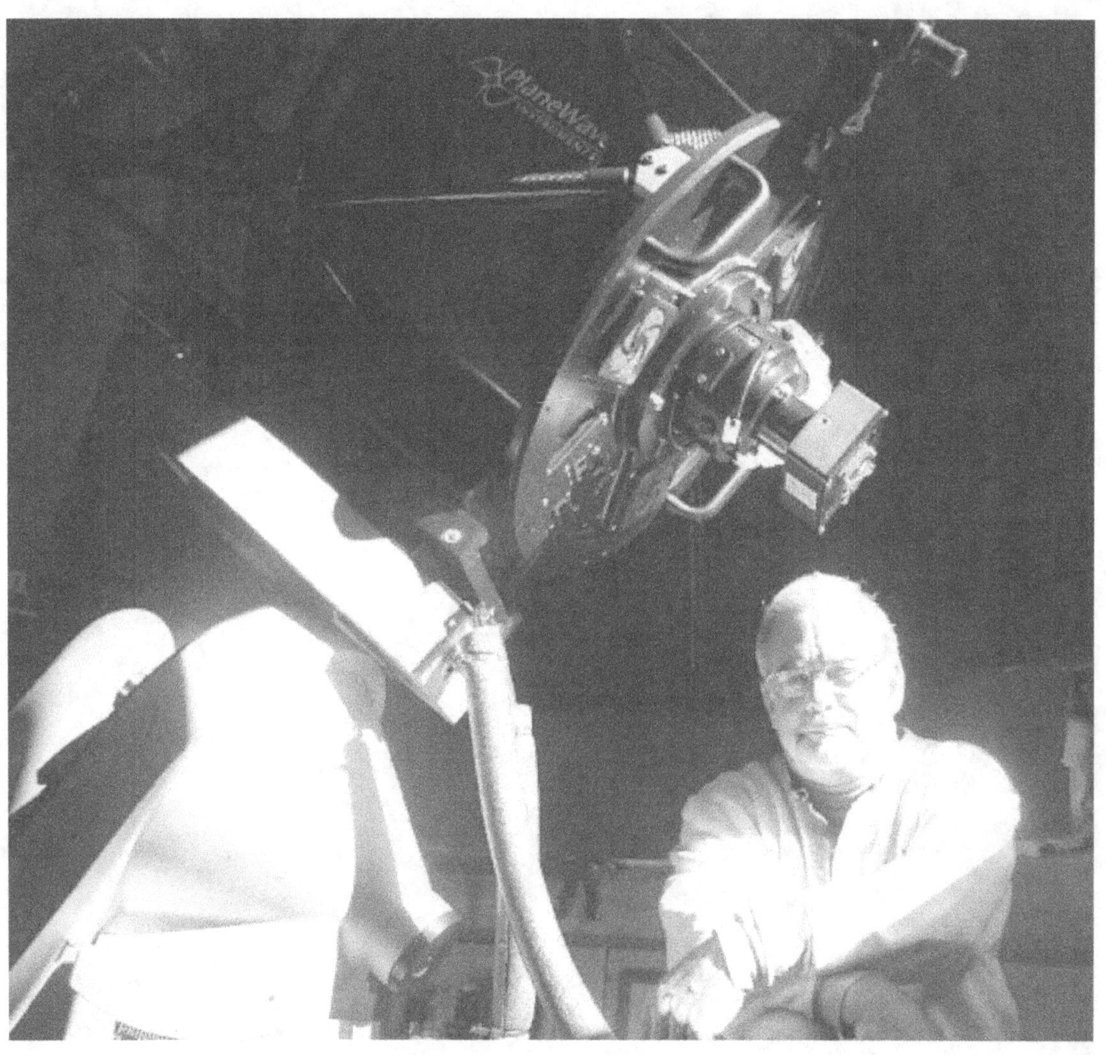

P. Clay Sherrod 2016
At the Arkansas Sky Observatories' 0.5m Astrograph
Petit Jean Mountain, IAU MPC H45

NOTES

NOTES

www.ingramcontent.com/pod-product-compliance
Lightning Source LLC
Chambersburg PA
CBHW081125170526
45165CB00008B/2559

* 9 7 8 1 3 8 7 5 3 7 1 9 8 *